Planetary Specters

Planetary Specters

Race, Migration, and Climate Change in the Twenty-First Century

Neel Ahuja

The University of North Carolina Press CHAPEL HILL

Set in Merope Basic by Westchester Publishing Services
Manufactured in the United States of America

The University of North Carolina Press has been a member of the
Green Press Initiative since 2003.

Library of Congress Cataloging-in-Publication Data available at
https://lccn.loc.gov/2021007907
ISBN 978-1-4696-6446-0 (cloth: alk. paper)
ISBN 978-1-4696-6447-7 (pbk.: alk. paper)
ISBN 978-1-4696-6448-4 (ebook)

Cover illustration: Girl walking along a broken embankment in Bangladesh,
2010 (© Jonas Bendiksen/Magnum Photos).

Contents

Illustrations

Planetary Specters

Introduction
The Specter of Insecurity

In public reports on "migration crises" in the Mediterranean and at the United States–Mexico border from 2015 to 2018, journalists and policy experts portrayed environmental changes as an underlying, hidden cause of mass migration. An article published in the UK newspaper the *Guardian* on October 30, 2018 is a case in point. Stating that "thousands of Central American migrants trudging through Mexico to the United States" have been affected by crop failures attributable to climate change, the article claims that the effects of environmental processes upon migration are "harder to grasp" than the more commonly reported bases of rapid displacement: "gang violence and extreme poverty." Displaying a photograph of the 2018 migrant caravan that journeyed from San Pedro Sula, Honduras to Tijuana, Mexico from October 12 to November 15,[1] the article proclaims, "The unseen driver behind the migrant caravan" is "climate change." Punning on the image of the open truck, photographed from the rear and filled with passengers from the caravan, the headline personifies climate change as the agent of displacement and configures the Honduran and Guatemalan migrants seeking asylum in the United States as part of a larger expected wave of "millions more" migrants who will flee to the United States.[2] In addition to quotations from two U.S.-based academics who argue that climate change exacerbates food insecurity in regions like western Honduras, which are experiencing large migrant outflows, the article includes the testimony of a single Mayan migrant from Honduras, Jesús Canan, who describes how drought "is forcing us to emigrate."[3]

Such reporting often includes images and descriptions of the hardships faced by migrants, as well as of the devastation wrought by increased atmospheric carbon concentrations, which accelerates global warming. However, it tends to ignore the language people use to explain political contexts in their home countries and the complexity of migration decisions. Organized largely through the efforts of the transnational immigrant rights group Pueblo Sin Fronteras, the migrant caravans attempted to articulate critiques of U.S. immigration policies, border detention, and the physical and sexual violence to which migrants are routinely subjected—both in

Migration

The unseen driver behind the migrant caravan: climate change

Oliver Milman *in New York,* **Emily Holden** *in Washington and* **David Agren** *in Huixtla, Mexico*

Tue 30 Oct 2018 01.30 EDT

13,758

▲ Honduran migrants taking part in a caravan heading to the US, walk alongside the road in Huixtla, Chiapas state, Mexico, on 24 October. Photograph: Johan Ordonez/AFP/Getty Images

Oliver Milman, Emily Holden, and David Agren, "The Unseen Driver behind the Migrant Caravan: Climate Change," *Guardian*, October 30, 2018, www.theguardian.com/world/2018/oct/30/migrant-caravan-causes-climate-change-central-america.

their home countries and on their northward journeys. Although they were viewed in the U.S. media as a sign of the large scale of mass migration (despite the fact that crossings at the southern border had actually dropped over the past two decades), the caravans were organized acts of political protest and mobilization to demonstrate the hypocrisy of U.S. asylum and immigration policies. Ignoring this performance context, the *Guardian*'s report suggests that asylum claims that migrants planned to submit at the U.S. border would exclude climate change as a deeper cause for their displacement: "Migrants don't often specifically mention 'climate change' as a motivating factor for leaving because the concept is so abstract and long term." By sidelining narratives of imminent danger, including targeted killings faced at home, stories emphasizing the climatic contributions to migration risk supplanting accounts of the political mobilization and legal strategies undertaken by migrants in relation to the different states they encounter on their journeys.

Rising atmospheric carbon concentrations are, indeed, intensifying ecological processes such as drought, desertification, extreme heat, sea level rise, coastal flooding, crop failure, and particulate pollution, which are destructive to human settlements and livelihoods. In 2017 alone, large flooding events displaced hundreds of thousands of people in locations ranging from Houston to Puerto Rico to Nigeria to India to Bangladesh, demonstrating that both environmental processes and failures of infrastructure to maintain dry land and to assist recovery are increasingly life-and-death matters.[4] However, climate migration discourse as articulated by journalists, security experts, politicians, and nongovernmental organizations (NGOs) often uses the complexity and widely distributed geographic nature of environmental processes to suggest that migrants represent the unwitting blowback of the destructive combustion and release of carbon wastes. Selectively relating interviews with displaced farmers and other rural workers who describe their experiences with the weather, such reporting jumps scale from localized weather to globalized climate, suggesting that environmental change increasingly constitutes the unacknowledged "push" factor behind migration decisions. The specter of future mass migration due to planetary ecological change thus configures migrants as embodiments of environmental processes. The figures of climate migrant and climate refugee in turn configure the imagery of liberal humanitarian concern for migrants as an expression of the Global North's overconsumption, evacuating the agency of migrants, removing focus from the proximate political and economic situations in their home contexts, and narrating their actions against the ghostly backdrop of the coming uninhabitability of their places of origin.

Critiquing Climate Migration Discourse

Planetary Specters: Race, Migration, and Climate Change in the Twenty-First Century presents a critique of the transnational discourse on climate migration. As the first book-length study in the humanities focusing on the intersection of race, migration, and climate change, it argues that the figures of the climate migrant and the climate refugee in public media and policy have, over the past decade, been constructed in a manner that obscures how the oil economy, economic inequalities generated by neoliberal economic and development policies, and forms of warfare and imperial intervention have been integrated into massive population movements that reflect capitalist regimes of racial disposability. Going beyond narrower legal and policy debates concerning proposed remedies for migrants affected by weather disasters, the book examines the invention and proliferation of climate migration discourse and imagery worldwide over the past three decades. The construction of multiple "migrant crises" in public media generates significant discussion of whether environmental factors related to the water cycle and food insecurity are driving migration. In the process, the questions of racism against refugees, changes in capitalist labor flows, and the imperial governance of northern borders are at times displaced in order to describe environmental forces as a sudden and severe generator of human mobility worldwide. In response to such crisis discourse, this book offers a focused study of changes in migration flows in Asia as a way to offer an alternative method for integrating economic, political, and environmental factors and analyzing their connections to forms of racial power. In the process, the book moves from a critique of planet-scale climate refugee discourse toward a more geographically focused study of how the systemic relationships between race and the oil economy have affected environmental destruction and migration flows originating in South and West Asia.

Discourses of climate migration highlight purported migration "hot spots" as sites of environmental degradation. Displaced groups and environments are in turn drawn into old-fashioned racial stereotyping about how the world's poor mismanage resources and engage in resource conflicts, a phenomenon that increasingly overlaps with newer forms of speculative prediction about how vulnerable groups will face coming climate destruction. Working against these twin forms of racial representation of climate change, *Planetary Specters* offers an alternative method for making sense of the connections among neoliberal economic policies, race, and environmental destruction in contemporary migration flows. This requires telling the story of how, beginning in the 1970s, the transnationalization of labor and trade was significantly influenced by the emergence of oil as the energy and financial basis of the global economy. From this vantage, the book argues that environmentally destructive shifts to inter-Asian manufacturing and labor recruitment were integral to what Cedric Robinson terms "racial capitalism"—the systemic generation of racial differences and inequalities through forms of capitalist reproduction and expansion.[5] From here, the book demonstrates how the stories of two of the key hot spots of purported climate migration—Bangladesh and Syria—can be retold by tracking how environmental processes relate to existing neoliberal labor migration pathways connecting agrarian peripheries to urban centers within these countries; transnational flows between West and South Asia; and transcontinental migrations from Asia to Europe and North America. Study of such migration pathways allows *Planetary Specters* to give an integrated account not only of how some of today's large migration corridors took shape but also of how state-led development policies and post-9/11 forms of warfare that generate migration are intertwined with oil-fueled changes in the world economy. Even as environmental processes are racialized within journalism, security policy, and emerging forms of green governance, the book highlights how the geographic expansion of capitalist production, new oil wealth in the Persian Gulf region, and the rise of new supply chains across Asia have created circuits of environmental and economic vulnerability that generate displacement—processes that tend to be masked in planet-wide accounts of climate migration.

To the extent that emerging climate migration discourses centered on Asia provide a dystopic vision of future displacement and resource conflict, they represent a contrast to celebratory visions of a New Asian Century, which have touted the gains of neoliberal finance across the continent over the past three decades.[6] This book examines how emerging visions of the underside of Asian development turn to some conventional scripts about the linkage between the agrarian poor and the border crisis. In the process, *Planetary Specters* works to challenge some racialized associations among rural peoples, disability, Islam, and war that emerge in journalism and security policy focusing on Syria, Bangladesh, and other sites of climate crisis discourse. Although climate migration discourse in these countries and worldwide has been propelled by liberal NGOs that claim a desire to reduce the injustices of either carbon emissions or immigration restrictions created by powerful states in the North, the humanitarian focus of such discourse recapitulates a history of racist colonial representations of Asia, Africa, and Latin America as zones of ecological degradation unable to achieve long-term development.

Meanwhile, the purported novelty of climate migration discourse in breeching divisions between nature and culture—the idea that it is a progressive arena of "posthuman" knowledge that integrates environmental and social analysis in new ways—is undercut by the ways it mobilizes racialized tropes of disability, Islamic insurgency, state collapse, and poverty. To the extent that the current increase in transnational migration is marked by the diversification of conditions of human mobility, it is also marked by a predictable retrenchment and militarization at the borders of many refugee-receiving countries. As such, the "migrant crisis" must be understood as "an unresolved racial crisis" that derives from neocolonial divisions in the international system.[7] This book focuses on such inequalities as they take shape in the Asia-Pacific region, where a number of important processes are concentrated: the transnationalization of the bulk of the world's manufacturing economy, the rise of unprecedented oil-funded development in the Persian Gulf states, the rapid changes in livelihoods due to rising waterways in the Pacific and Indian Ocean regions, the world's largest migration flows from rural to urban areas and from poor states to the Gulf oil producers, and the militarized surveillance of ethnicity and religious affiliation across West and South Asia. If climate change discourse tends to abstract such geographic itineraries of carbon-fueled capitalism by emphasizing that "anthropogenic" or human-made emissions broadly cause environmental devastation, *Planetary Specters* attempts to analyze how unequal distribution of the benefits and costs of carbon pollution systematically produces racialized border crises. A critical analysis of migration and climate change requires attention to how the histories of race and capitalism intertwine to produce population movements, trade linkages, and forms of environmental destruction at large scales. By highlighting the interrelation among securitization, neoliberalism, oil production, and environmental change, this book presents interconnected geographies of displacement as an alternative to simplistic efforts to label particular migrants as climate refugees or particular migration routes as climate migration pathways.

Climate Change as a Border Crisis

One of the basic lessons we learn from reading scholarship in the field of critical refugee studies is that the view of migration as a problem or crisis for a nation-state in which migrants arrive tends to mask forces that displace people from their home countries. In much of the public journalism and policy discourse on migration in North America, Europe, Australia, New Zealand, and other high-income countries, it is the scale of migration and the purported problems of migrants' economic needs, population pressures, legal status, racial or cultural differences, and assimilation processes that have historically preoccupied journalists and politicians focusing on migration. The economic demand that these receiving countries create for migrants to do low-wage labor to sustain their economies, and the forms of militarism and development they impose on parts of the world that send migrants, often take a back seat to crisis discourse about their arrival—that is, to narratives and images that suggest migrants are swarming the borders and threatening to change the economic and social bases of the nation-state. Whether sympathetic to or dismissive of migrants' struggles, these crisis discourses figure the transnationally mobile body as a problem in itself rather than an effect of broader processes of economic inequality, war, and social change. As a field of study and writing informed by critical social theories, migrant narratives, and antiracist and immigrant rights movements, critical migration studies asks us to reframe the issue by focusing on the systemic nature of migration.

Public discourse on migration in the past decade has prominently figured global warming as a key factor driving people's decision to flee their home countries. In the wake of a sustained political attack against international emissions mitigation agreements and climate science by fossil fuel interests and governments that back them, climate migration discourse appears to introduce a new avenue in mainstream journalism to politicize the impending effects of climate change. In the absence of a working international project to curb carbon emissions, the figures of the climate migrant and the climate refugee appear to help make appeals through national security apparatuses to create new climate policies. In this vein, news articles, policy papers, security think-tank reports, and documentaries focusing on the growing human costs of environmental change create a new avenue for politicians and advocacy groups to argue for individual countries or regional entities to take unilateral security action to minimize the costs of climate change. Does this represent a new and progressive change in the manner in which migrants are represented, especially in the wealthy countries that have the greatest share in the global contributions to atmospheric carbon emissions? Emerging environmental discourse on migration is especially notable in journalism in countries in the Global North—including the United States, the United Kingdom, and Germany—which have long been primary immigrant destinations and major polluters. In these locations, where there are active right-wing political movements that maintain xenophobic focus on immigration's purported negative cultural and economic effects, climate change is configured as a security concern signaled by mass population movement.

Questioning emerging security discourses about the links between climate change and migration is necessary to reframe the manner in which militarized northern borders contribute to the current sense of a "migration crisis." Rather than migrants themselves, the forces of capitalist enterprise and military intervention are the most prominent border-crossing agents today, forces that generate growing numbers of southern migrants as their unacknowledged supplement. As Harsha Walia of the Canadian immigration activist group No One Is Illegal succinctly puts it in *Undoing Border Imperialism*,

> Capital, and the transnationalization of its production and consumption, is freely mobile across borders, while the people displaced as a

> consequence of the ravages of neoliberalism and imperialism are constructed as demographic threats and experience limited mobility. Less than five percent of the world's migrants and refugees come to North America. When they do, they face armed border guards, indefinite detention in prisons, dangerous and low-wage working conditions, minimal access to social services, discrimination and dehumanization, and the constant threat of deportation. Western states therefore are undoubtedly implicated in displacement and migration: their policies dispossess people and force them to move, and subsequently deny any semblance of livelihood and dignity to those who can get through their borders.[8]

We need look no further than the recent crisis narratives about the migrant caravans at the United States–Mexico border and the Mediterranean crossings into Europe as evidence of this imbalance in the public discourse on migration. Honduran migrants are configured as security threats to the United States rather than as groups facing displacement in the wake of the U.S.-supported 2009 coup removing Manuel Zelaya from the presidency. The ensuing right-wing paramilitary violence that many migrants discuss as a reason to emigrate was emboldened by the National Party that took control subsequent to the coup.[9] Similarly, Syrian refugees are portrayed as unassimilable in German media, sentinels of social ills of sexism and Islamism.[10] In such crisis reporting, familiar rhetorics of external threat and post-9/11 fears of terrorism highlight how the migrant body is configured as the problem, rather than the systems of extractive capital, labor exploitation, decentralized warfare, and militarized borders that accelerate displacement and vulnerability.

Furthermore, the crisis discourse about climate migration emerges in a context in which environmental factors are increasingly viewed as accelerating border-crossing threats to international security. This requires that we consider some perhaps unexpected linkages among race, religion, and environment; the figure of the climate migrant as a security risk in public policy reports appears to link climate change to other security figures, including the figure of the terrorist. The launch of post-9/11 wars worked to entrench a model of militarized security that went beyond the ostensible goal of responding to the rise of Islamist insurgency. As scholars in critical security studies have argued, northern states in the wake of 9/11 invoked a broad systems-oriented approach to security that combined stereotyped views of Islam's threat to the state system with broader attempts to surveil environmental and human networks, out of which threat potentials are understood to constantly emerge.[11] Environmental phenomena may seem unrelated to the stereotyped construction of the terrorist, but from the vantage of security officials, they are parallel in their capacity to disrupt the smooth functioning of state and economic systems. At the same time, current ideas about security planning emerge within a longer context in which the supposed postsocialist transitions following the fall of the Berlin Wall and the dissolution of the Soviet Union rendered Islam and socialism as radicalized signs of threats to U.S. and European liberalism's ways of conceptualizing human futurity.[12] From here, the failures of development in the South—environmental degradation, population pressures, failures of social provision—appear as potential drivers of "extremism." Since such security discourses disavow race's central role in the unequal formation of the international system, they reify precepts that environmental change may be driving civilizational differences, suggested by a division between Islamist and secular views of governance. Put differently, the postracialism of transnational climate security discourse unites a posthumanist outlook on environmental risk with a postsecular conception of the international. In the process, the northern security state's xenophobic views of Islam combine a complex set of environmental factors into visions of human vulnerability, ensuring that representations of individual migrants as well as entire nations remain linked to fears of Islam as a threat to democratic life.[13]

In emerging security thinking, integrating forms of surveillance for human and environmental phenomena generates a fantasy of "human security" from the multiple depredations of intentional or accidental vulnerability. In this context, racism involves not just a set of stereotypes attributed to the body, such as skin color, or cultural attributes, such as religious dress; it involves a dimension of time: "migrant crisis" emerges from the background of everyday life into the spectacle of sudden disaster. It is precisely the fact that environmental processes such as climate change appear to come from long-term, purportedly unintentional processes beyond human control that makes their capacity for racial spectacle salient. As environmental changes suddenly erupt into the lived experience of a population and threaten a sense of stability in time and space, their potential to virally unleash unpredictable social or behavioral responses configures a need for securitization against unfolding risks. If climate change undermines the normal social bonds, the story goes, an opening is made for Islamists, rogues, criminals, and other racially characterized figures of security risk to emerge. Climate change could be mitigated at the outset, but failing that, its effects will have to be policed. Islamophobic invocations of terrorist threat appear on the margins of climate migration discourse, as U.S. and European policy analysts and military advisors regularly note the fact that many places affected by desertification and sea level rise are located in Muslim-majority states like Syria and Bangladesh.

One could reasonably argue that the enhanced border controls that came with the coronavirus pandemic in 2020 might put on hold some emerging security discourses emphasizing climate change's role in migration and conflict. However, current trends in both the acceleration of global warming and the continuation of human displacement point to the continuing significance of such ideas in the coming years. In the United States, President Joe Biden's pre-inauguration climate plan drew on climate security discourse, emphasizing that climate change is a "threat multiplier" that requires addressing "water scarcity, increased risks of conflict, impacts on state fragility, and the security implications of resulting large-scale migrations." Such implications include, according to the plan, "deteriorating economic conditions" which "could increase piracy and terrorist activity, requiring a US military response."[14] Although the COVID-19 control measures adopted in spring 2020 have slowed economic activity worldwide, at the time of this writing usage trends indicate that carbon emissions are not likely to see a long-term decline. The study that provides the most comprehensive data yet on the decline in energy use related to pandemic lockdowns does show a notable short-term dip in fossil fuel use and emissions, bringing current emissions closer to those of 2010. But the growth forecasts suggest a likely increase.[15] There has not been a dramatic shift in the wealthy countries away from fossil fuel dependency; assuming that plans for nation-states to reopen economies without first transforming energy use stay in place, we can expect that the pandemic and the emerging economic recession or depression will slow the growth rate of fossil fuel use in the short term but not reduce total carbon emissions substantially enough to meet international emission reduction targets. The likely return to carbon emission growth is especially urgent, as according to prepandemic estimates, energy use—fueled mainly by surging oil and gas consumption—was projected to increase worldwide by 28 percent by 2040.[16] Climate security measures will thus likely remain aimed at containing the aftereffects of emissions rather than addressing the problem through concerted efforts to rein in emissions. By all reputable accounts, rising global temperatures will increase the incidence of heat-related deaths, disease, food insecurity, and destructive weather, affecting poor countries most severely due to already overstressed infrastructures.[17] Meanwhile, the international economic inequalities that may influence migration decisions have only intensified during the pandemic. The massive flows of people domestically in India and across borders in Latin America following the onset of COVID-19 lockdowns indicate that new controls on mobility have not simply led to a pause in migration but have in some cases exacerbated displacement and, indeed, made migrants more subject to forms of stigma, violence, economic desperation, and health risk.[18]

Critical Refugee Studies and the Environmental Migrant

Even though in this book I am critical of some invocations of "climate migration" by journalists and security experts, I don't mean to dismiss discussions of how environmental processes constitute one important factor contributing to human mobility. *Planetary Specters* argues that environmental injustices must be understood as components of longer processes of colonialism and racial disposability generated by extractive capitalist development. The fossil fuel–based energy system that emerged from colonial histories of extraction is a critical facet of present-day environmental racism. Certainly the rise in flooding disasters in the early twenty-first century is a signal that temporary displacement after inundation is becoming normalized as a condition of human habitation across many coastal and island regions, as well as in places where poor infrastructure exacerbates the problems of growing rainfall or rising waterways.[19] There are reasons that activists and politicians from small island states in the Pacific Ocean, Indian Ocean, and Caribbean—most prominently those from Kiribati, Tuvalu, the Marshall Islands, and Fiji in the Pacific and from the Maldives in the Indian Ocean—have been forceful in their proclamations that climate change will permanently displace Indigenous groups from their places of origin and thus from their cultural and ecological heritage.

Crisis discourse focusing on climate migration has emerged in tandem with a recent turn in migration studies to considering environmental causes of migration, particularly given the interest among development scholars in drought and famine. However, this turn to discussions of environmental migration in the field should not be understood as a new addition of environmental factors to a field that has historically focused on economic factors. In fact, the nineteenth-century origins of sociological investigation of migration included significant focus on environmental factors, which were subsequently critiqued in the post–World War II era because of their determinist views of social action. Reorienting the field around economic questions, several decades of migration studies literature in the late twentieth century considered ecological factors marginally if at all. The rise of refugee studies during this period, responding to the postwar elaboration of international asylum law, tended to confirm that "forced migration" was understood in political terms. States that targeted minorities with discriminatory laws and citizenship regimes were seen as the primary generator of the need for humanitarian refuge. Contextual considerations such as economic and environmental factors were excluded from the definitions of refugees. As the field has expanded with critical methods focused on race, empire, class, gender, and sexuality, the recent invocations of environmental migration among anthropologists, sociologists, and geographers have become controversial. Scholars have commonly noted the alarmist tone and speculative outlook of some of the extant studies on the topic, as I discuss throughout this book.[20] Emerging work in critical refugee studies attempts to dislodge the normative legal associations with the category of the refugee, creating a new opening for research that breaches the boundaries between political, economic, and environmental categories as well as the binary division of migrants from refugees.[21] Thus at present, there is an opening for critical refugee studies to chart a new path to analyze the complex environmental, economic, and political determinants of migration.

As I discuss in chapter 1, the idea that environmental crises are systematically driving accelerating numbers of migrants across borders has only been a significant topic of study among development, migration, security, and environmental experts over the past thirty years. Public awareness of concepts like climate refugees or environmental migration has had an even shorter shelf life of about one decade, as failure to reach international agreements on climate change coincided with the economic shocks, new military interventions, and regional wars that came after 2008. Since then, in the midst of a rapid uptick in transnational migration, a cottage industry of pundits, think tanks, and nonprofits has developed a growing literature on climate-influenced displacement and climate insecurity issues, focusing in large part on forms of international conflict that might emerge due to rapid environmental shifts or resource shortages. Despite the relative lack of academic research on links between migration and climate change, concern about global warming as a driving force in the current displacements is growing, and crisis thinking is quickly generating depictions of Black and Brown migrants from Africa and Asia fleeing shrinking zones of habitability. In much of this reporting, there is a familiar form of racialized humanitarian governmental imagination coming into focus, with the representation of mass displacement constituting purported climate refugees as objects of intervention assistance and management by first-world NGOs and international agencies.[22]

As such, critical migration and refugee studies have a role to play in rethinking standard accounts of climate migration and conflict. Even with public invocations of forced environmental migration as a violation of human rights and Indigenous sovereignty, the subjection of climate-affected peoples to state securitization and humanitarian intervention carries complex risks. On what bases do contemporary humanitarian or security discourses applied to ecologically transitional areas create racialized knowledge, technologies, and interventions that transform the lives of those affected by climate change? Do those discourses offer pathways for addressing the forces that generate displacement as a form of structural racism, reproducing what Ruth Gilmore defines as "state-sanctioned or extralegal group differentiated vulnerability to premature death"?[23] Or are we witnessing how climate security discourse is helping to produce new forms of racialization in speculative constructions of at-risk groups whose homes and homelands are configured as uninhabitable (and thus disposable) in a warming world?

Early signs suggest that climate securitization only exacerbates racial divisions in life outcomes, which are, in part, expressed through the large increases in internal and transborder migration globally. Since U.S. imperial militarism, the collapse of the socialist states, and neoliberal transnationalization of industrial production and labor—the united forces of a purportedly postsocialist world order—have generated increased transborder migration over the past three decades, significant changes in the geographies of human mobility have emerged. Today, more and more migrants from the Global South seek safety in other parts of the south, a development that in itself generates alarm among northern security experts who see the potential for a spillover effect whereby the most vulnerable southern countries will undermine the state structures of stronger ones. This view is itself reinforced by the fact that south–south migrants regularly move on to the next step of building resources for the difficult and potentially deadly subsequent journeys they may take toward the militarized borders of Europe, the United States, or Australia. Such multiple migrations are a source of fear for security experts and right-wing politicians in the North. The Persian Gulf states today are the site of massive diasporas from South and Southeast Asia (especially from India, Pakistan, Bangladesh, Sri Lanka, and the Philippines), and major migrant flows depicted in the news media move large numbers of people from sub-Saharan Africa to the Mediterranean;

from Central America to Mexico; and from many coastal, island, and rural locations to urban sites across Asia. The post-9/11 wars launched by the United States and increasingly prosecuted by a variety of state and non-state actors have displaced millions across the borders joining Iran, Iraq, Syria, Lebanon, Turkey, and the Kurdish-majority regions therein.

With these routes and others, current estimates of transnationally displaced people reach above sixty-five million. Environmental forces today are intersecting with unequal regimes of rights and citizenship that affect the daily lives of those who migrate. Taking note of how sea level rise is affecting small island states in the Pacific Ocean, Walia suggests its growing influence on transborder migration: "Tuvalu is one of dozens of low-lying Pacific Island nations threatened with total submersion as climate change and global warming cause ocean levels to rise drastically. . . . Over one-fifth of Tuvaluans have already been forced to flee their country, many to poor neighboring islands such as Fiji, and others to New Zealand. Despite having the world's highest emission per capita at 19.6 tons of carbon dioxide per person, Tuvalu's other neighbor, Australia, has so far refused to accept Tuvaluans as climate refugees. Border imperialism again denies justice to migrants who are its own casualties."[24] Walia's focus on these small island states recognizes that they are on the front lines of some of the most sudden and intractable impacts of carbon pollution. Small islands that are losing landmass to sea level rise are sites of the most immediate displacements due to climate change. As the islands that are receding most quickly displace residents to larger nearby islands, those areas in turn are affected by a combination of population pressure, land degradation, and sea level rise. In response, climate justice activists from small islands in the Pacific have attempted to raise global awareness of how climate change threatens individuals' lifeworlds through these processes of displacement, removing people from heritage lands and alienating them from communities. Taking note of the casual atmospheric violence meted out by large polluters against smaller, more climate-affected states, Walia situates climate change as one of the forces of displacement generated by the systematic inequalities of imperialism and capitalism that remain largely out of the purview of political debates on migration. Despite the undeniable pressures that these small countries face amidst loss of land base, the legal architecture of border exclusions prevents redress.

As I discuss further in chapter 1, the extant legal petitions by migrants for asylum based on climate change have failed to yield any individual redress, even as the scale of migration has swelled. One key challenge for critical studies of migration in this context is to develop methods of integrated geopolitical, social, and environmental analysis that can interrupt the crisis discourses that alternate between humanitarian concern for migrants and xenophobic invocations of economic chaos, terrorism, crime, and loss of purportedly secular national culture.

Racial Capitalism, Colonial Energy, and the Deep History of Migration

One strategy that scholars in critical race and ethnic studies, critical refugee studies, and postcolonial studies have deployed to challenge racialized spectacles of border crisis is to reframe the present crises within a longer time span. This helps us understand how the changes in land use and exchange generated by European franchise colonialism (the establishment by countries including England, France, Spain, Portugal, and the Netherlands of commercial empires, including formal colonial possessions in Asia, Africa, and the Americas), settler colonialism (the establishment of breakaway white-majority colonial states in the United States, Canada, New Zealand, Australia, South Africa, and elsewhere), and the related development of capitalist property systems have created widespread conditions for human displacement and racial disposability.[25] Race historically emerged as an important regime of colonial statecraft because it managed human population demographics and land use in ways that reproduced a profitable mix of commoditizable land and labor; racial regimes of colonial states have thus exhibited some flexibility in order to manage human difference in ways required to make land, labor, or natural resources exploitable.[26] Viewing present ecological disasters including climate change as an outgrowth of colonial forms of labor, production, and energy use helps us understand something about how race, as a flexible regime of colonial power and profit, and racism, as the structured management of group vulnerability to premature death, have shaped ecologies of migration. The moves by states with a history of white supremacy encoded in law to formally declare racial equality in the aftermath of World War II, decolonization struggles, and the antiracist movements of the 1960s have to date failed to stem the economic conditions that unequally render large swaths of the world's agrarian poor subject to the depredations of the global economy's financial, environmental, and political violences.

By the 1970s, fossil fuel development exacerbated global wealth inequalities and committed billions of Black and Brown people across Asia, Africa, and the Americas to dependencies on U.S.-led financial networks. The environmental and economic nexus of racial capitalism, which I discuss in detail in chapter 2, also breached the borders of the wealthier countries, with Indigenous lands devastated by oil extraction and sea level rise, and large, racially marginalized labor diasporas forming across the former colonial powers of Europe, the settler colonies spanning the Pacific Rim (the United States, Canada, New Zealand, Australia, and South Africa), and the newly rich, formerly colonized petrostates of the Persian Gulf. These neocolonial relations—which exacerbated some preexisting international wealth divides created by colonial labor and trade systems—emerged at the very moment that a series of economic crises suggested the possible decline of European and U.S. financial power. But the oil economy helped to float U.S. government debt (in part incurred through the massive expenditures of the Vietnam War) and to sustain the centrality of northern financial and corporate power in transnational capitalism. Thinking about oil-driven climate change as an outgrowth of the emergence of a planetary system of racial capitalism—which reshaped land use and based profit on the development of racially stratified labor systems—is necessary for a historical understanding of environmental disasters that appear as acute, contemporary displacement events. In this book, the critique of representations of migration as a climate change disaster is accompanied by a historical examination of how fossil fuels, especially oil, contributed to racially unequal conditions of displacement—even if large-scale extraction of fossil fuels never fully destroyed the diversity of human relationships to environments.[27]

Although there are different historical reference points for the processes that originated the current environmental crises—the formation of capitalism and the establishment of property in land in the sixteenth century, the expansion of slavery and the plantation economy in the eighteenth century, the rise of the coal economy during the height of nineteenth-century colonial land grabs, and the accelerated use of oil in the mid-twentieth century—scholarship in colonial environmental history, world systems analysis, and green Marxist thought contends that current environmental problems such as rising global temperatures, desertification, erosion, soil degradation, wildfires, and increased storm activity are products of longer developments in the extractive relationship between human societies and their environments. These forces have historically embedded the sustenance and reproduction of national economies in processes that deplete ecological capacities, displace people from traditional homes and lands, and render rural people dependent on long-distance markets.[28] For Indigenous scholars focusing on the lengthy histories of settler colonialism in the Pacific and the Americas, European settlement produced massive ecological shifts. By reducing open common lands, introducing diseases, reducing cultivated forests and native grasslands, and introducing invasive species, direct colonial warfare was supplemented by massive forms of ecological change that accelerated colonial genocide.[29]

The imposition of a colonial land system and the violent depopulation of Indigenous societies by such settler states as the United States, South Africa, Brazil, and Australia emerged over several centuries. During the consolidation of the state structures of these settler colonies—each with racially differentiated citizenship hierarchies—the mixture of commoditized human labor with renewable and nonrenewable energies powered the growth of capitalist relations across geographies of dispossession and concomitant migration. Building on technologies of displacement emerging from the enclosure of the commons in England and western Europe, the rise of the Caribbean plantation complex and the Atlantic trade in enslaved peoples from the seventeenth through the nineteenth centuries ushered in a form of transcontinental interconnection that helped to geographically expand both the capitalist regimes of property in land and the fungibility of labor and life as the bases of colonial profit and demographic expansion of white settlement. As plantation expansion coincided with deforestation, which allowed for expanded shipbuilding in Europe as well as heating and cooking fuel in the colonies, European settlers developed forms of colonial land use that sustained the geographic expansion of capitalism across the Atlantic. This included the development of geological surveys to help map settler colonial expansion for both agricultural and mining capacities, a history that embeds contemporary geological theories about climate change in longer histories of colonial knowledge production and land reorganization.[30] Plantation agriculture set in motion massive forms of human labor accumulation and migration, including the forced movement of over ten million enslaved people from West Africa to the U.S. South, the Caribbean, Brazil, and other slave states in the Americas, followed by the importation of several million Asian indentured workers in the nineteenth and early twentieth centuries. The centrality of anti-Black racism to this system of labor exploitation created the basis not only for other groups to be subjected to coercive labor regimes, but for the elaboration of mythologies that the sale of one's labor provided a type of exercise in human freedom that moved one away from the condition of enslavement.[31] For this reason, the system of slavery must be understood as part of the larger processes of generating myths of the universality and independence of the figure of the human, despite forms of racial sorting that emphasized the fungibility and affectability of large populations in Africa, Asia, and the Americas.[32] Plantation expansion coincided with the concentration of small landholdings by the British in India, allowing a similar process of displacement and labor appropriation that advanced distance trades in key colonial agricultural commodities. The development of this transcontinental economy involved the alienation of large numbers of colonized peoples from small-scale agrarian production and their relocation into larger-scale plantations that, in turn, accelerated the rise of manufacturing and the rise of new trading linkages in the North Atlantic and across Europe. As such, the colonial concentration of labor in cities forms a key part of the process of developing racial capitalism's capacities to conduct production and trade across expansive imperial geographies.[33]

The rise of racial capitalism out of systems of colonial extraction required the development of large-scale energy systems. The plantation complex involved the admixture of human and nonhuman animal labor with solar power (in the photosynthesis of crops) and wind power (in transoceanic shipping) for the distance trade of monoculture commodities like sugar, rice, tobacco, cotton, spices, and indigo. The twentieth-century rise of fossil fuels changed the relationship between geographies of migration and geographies of production and exchange. In locations in the United States and Europe, where the steam engine brought coal use beyond its traditional application for heating and cooking, manufacturing processes accelerated the rate at which basic agricultural commodities could be transformed into textiles and other finished goods for export.[34] The expansion in carbon emissions due to coal use inaugurated the fast forms of global warming that we are witnessing today. Later, the rise of oil production competed for agricultural lands yet required fewer human laborers to extract profit.[35] Coal and oil use increased the capacities for distance shipping and manufacturing, in turn redistributing the imperative for basic commodities to be grown or extracted in more dispersed locations worldwide. While the significance of migrant agrarian labor in maintaining both food economies and basic commodities for manufacturing grew throughout these transformations, some laborers were rendered surplus to production requirements as a result of increased mechanization and rapid transport. Meanwhile, control over land by extractive industries and large-scale farming, housing, and industrial operations remained geographically uneven, leading to growth in migration from peripheral small farms to both large rural landholdings and urban centers of trade and manufacture. If the plantation complex reinforced the reliance of a geographically expanding industrial economy system on transcontinental migration and shipping, the rise of fossil fuels allowed increased mechanization of extraction and production, sped the concentration of colonial land bases into large holdings, and grew divisions between agrarian and industrial workers worldwide. Thus, the connections between colonial geographic expansion, the change in the land base prompted by plantations and mining, the growth in manufacturing and distance trades created by fossil fuels, and the capitalist trades in both manufactured goods and laborers were part of a large-scale set of processes creating new migration and trade connections across continents and between rural and urban centers.

The systemic expansion of capital across the sixteenth through the twenty-first centuries thus yoked growth in fossil fuels to the privatization of land and the alienation of labor in ways that made migration central to the reproduction of capitalist systems. As national, religious, and ethnic markers continued to differentiate migrant labor forces within these systems, the racial character of capitalism remained crucial to the potential for managing dissent and preserving the dominance of oil-based transport and finance in maintaining the geographic reach of trade. As such, to the extent that an extractive capitalist influence on the environment increased over five centuries of colonial history, it also systematically generated human migration as a concomitant process to climate change, species extinction, and other forms of environmental destruction.

Inter-Asian Migration Ecologies

The planetary effects of this long-term colonial process of alienating people from heritage lands have been sped up by the combination of energy and trade transitions that are unique to the late twentieth and early twenty-first centuries. The capitalist processes that have given rise to anthropogenic changes in climate have specific geographic histories that are significant for understanding not only environmental change but also itineraries of racial power and struggles over labor and resources. I argue in *Planetary Specters* that the role of countries in Asia and the Pacific in the system of racial capitalism is significant to understanding the economic and environmental nexus of migration today. At the same time, the lessons we learn from migration routes in the Asia-Pacific offer larger lessons about how racial capitalism relates to environmental destruction, displacement, and governance worldwide.

Mass fossil fuel use has played a role in both industrial development in the North since the nineteenth century and the shift in the twentieth century from the United States and Europe to Asia as the manufacturing and logistics center of the world economy. Although the rise of coal and steam power first energized manufacturing in the United States and England in ways that allowed each of those countries to assume a central role in nineteenth- and early twentieth-century transnational trade (with its destructive carbon emissions), coal power eventually became the basis of national electric grids around the world and today still constitutes 38 percent of the world energy mix. At the same time, the expansion of the oil economy during the interwar period suddenly shifted the geographic basis of the fossil fuel economy to several key areas of extraction, including Texas, Nigeria, Venezuela, India, and, most centrally, the Persian Gulf states. This coincided with the end of the gold standard and the establishment of the role of the U.S. dollar in oil and arms trades, reconfiguring how national energy provision ended up requiring international exchange in dollars. In turn, a number of oil-rich former colonies attempted to establish more control over oil as a key economic resource for national development. The rise of the Organization of the Petroleum Exporting Countries, which I discuss further in chapter 2, is one notable aspect of this history.

Although the aspiration for economic independence from U.S. financial power in formerly colonized states was part of the history of geopolitical conflict over oil, the history of oil as a key energy commodity fueling new regimes of development and trade tended instead toward international economic dependencies on U.S. finance that made the international system increasingly unequal and unstable for the poorest formerly colonized regions and populations.[36] Oil fueled the shipping regimes that allowed the rise of neoliberalism, an economic rationality in which states relied increasingly on dispersed, privatized economic activity to replace organized state systems for human provision and welfare. Even as some states—most notably Iran, Iraq, the United Arab Emirates, Kuwait, and Saudi Arabia—attained unprecedented GDP growth based on the new wealth from oil extraction, these states failed to widely distribute the economic gains across class divisions or to fund regional economic collaboration and stabilization among southern states. As organized challenges to U.S. and Soviet empire such as the third world movement floundered in this increasingly cutthroat international economic context, growing dependency on imported fuel and food in the poorer nations led to displacement of agrarian subsistence economies that provided local economic stability. People across the Global South became more subject to market instabilities. The spikes in oil and food prices that cyclically emerged during the era of neoliberalism energized circuits of national and transnational migration.

Yet the effects of these processes also helped to concentrate capital in a new set of manufacturing and financial centers across East and Southeast Asia, with much of the world's textile, automotive, and electronics production chains now operating through Asian contractors and factories. To staff these new industries, large numbers of workers migrated internally, and in some countries, new low-tax, export-oriented economic zones were established to encourage multinationals to set up shop. Oil made this system of flexible labor and production possible, as transnational corporations took advantage of uneven national labor and environmental regulations to profit from surplus laboring populations in Asia. Although China remains central to this process in terms of its massive factories and huge rural-to-urban migration pathways coincident with its absorption of industrial capacity, smaller countries with cheaper operating costs (Bangladesh, Vietnam, the Philippines) have high-growth export sectors that feed into longer transcontinental commodity chains. As more people in these countries have been displaced from agrarian lands and local trades, they have been increasingly subject to transnational market forces that encourage them to migrate to cities or other countries and to remit a portion of their wages to their home economies.

These processes have been compounded by massive environmental changes wrought by the use of oil and coal for manufacturing and shipping, especially in the water system, which is so crucial to the livelihoods and settlements of the economies of rural and coastal areas. Although there are many environmental disasters that characterize the era of neoliberalism and track closely with expanded oil use, I focus in this book on interlinked climatic, economic, and geopolitical changes that have fueled migration in South and West Asia. The massive outmigration of agrarian workers from rural and coastal areas of South Asia has coincided with the expansion of the export-oriented manufacturing base in South Asian cities and the departure of a huge diaspora to the rapidly developing Persian Gulf states. The migrations in the eastern Mediterranean during the ongoing regional wars emerge as widespread violence compounding economic crisis has fueled new migration routes. In these locations generating significant twenty-first century migration from West and South Asia, we witness migration pathways where climate-driven changes to water systems (including drought, flooding, desertification, increased salinity, and loss of drinking water) overlap with major economic and geopolitical forces that conduct human resettlement. These interlocking forces include oil-fueled development in the rich Persian Gulf states, which seek South Asian migrant labor as a workforce that is alienated from regional political struggles; the expansion of Pacific shipping economies, which ensure the rapid delivery of oil from West Asia and finished goods from across the continent; and the expansion of arms trade and armed interventions in West Asia, which link imperial forms of securitization and militarization with ongoing transformations in agriculture in the region. Environmental changes to the water cycle contribute to pressures for migration in this context even as migration is increasingly managed by racialized forms of crisis governance.

A Posthuman Politics of Scale

The period of the rise of the oil economy since the 1970s, which witnessed intensified human displacement globally, coincided with several distinct currents of environmental thought through which environmentalists attempted to grasp the transitions at hand. As I discuss in detail in chapter 1, northern NGOs in the 1980s and 1990s developed a significant focus on southern overpopulation as a problem that appeared to drive transborder environmental crises; one of the main concerns at the time was how the growing population led poor people to destroy forests for agriculture and engage in forms of subsistence farming that degraded the soil and the water table. Drawing conceptually on the ideas regarding ecological carrying capacity developed by Thomas Malthus, such a focus tended to concentrate on high rates of women's reproduction in the Global South rather than structural economic and political changes as a significant driver of environmental change. This type of gendered and racialized thinking about population was closely linked to sustainable development discourses that worked to entangle conservation in formerly colonized nation-states with economic development schemes such as ecotourism and, at times, with family planning initiatives to control reproduction rates. By 2000, amidst growing international concern about rising global temperatures, environmentalists and scholars also began to debate the concept of the "Anthropocene"—the phase of Earth history in which the human species purportedly determines the geophysical structure of the planet.[37] Although the concept was initially generated in the context of attempts to formalize a new epoch in Earth's history among geologists, the idea has gained wide attention beyond that discipline within academia and journalism.

How do climate migration narratives interact with these currents of environmental thought? Recent scholarship in the environmental humanities highlights how climate change challenges conventional understandings of history, literature, and politics, arguing that fossil fuel energy use has, since the industrial revolution, radically reshaped planetary life.[38] In broader academic conversations beyond the humanities, debates among geologists, geographers, and social scientists over the concept have taken a strong ethical cast. Critics of the Anthropocene concept point to the outsize responsibility that the United States and Britain have for historical carbon emissions, which are unequally affecting the world's poor and the Global South. In turn, scholars working on approaches to the environment in Black studies, postcolonial studies, and Indigenous studies, including Kathryn Yusoff, Amitav Ghosh, and Kyle Powys Whyte, have been critical of abstracting the "human" (*anthropos*) influence on climate; they argue convincingly that attention to the violences of racial slavery, class exploitation, colonial land use, unequal environmental pollution burdens, and the very field of geology as a colonial science have shaped the interests and actors that have made human impact on the climate a field of deep inequalities.[39] In the process, a wide array of scholars of race, colonialism, and capitalism have invented a series of terms that propose to specify more clearly the responsible parties or processes: Capitalocene, Plantationocene, Eurocene.[40] The resulting discussions of the Anthropocene counternarratives fit within a broader critical race and ethnic studies critique of the postracialism of recent posthumanist and new materialist trends in the humanities.[41]

For the purposes of this book, I articulate another way that we need to understand the relevance of critical race analysis to Anthropocene discourse. If we situate the rise of Anthropocene narratives and counternarratives within a recent history of environmental representation in which conservative, neo-Malthusian ideas about population in the Global South have been widespread in expert discourse on climate change, we can glimpse how the abstraction of human responsibility for environmental damage under the banner of the Anthropocene may have initially appeared as a progressive transcendence of earlier, stereotyped visions of poor, agrarian Black and Brown people overpopulating and destroying local ecologies. The Anthropocene story is a story that brought the North's overconsumption into the narrative of degradation, even if it failed initially to substantively detail unequal pollution burdens and benefits. Critiques of the Anthropocene concept might benefit from specifying this immediate pre-history by combining three neocolonial dynamics: a massive increase in the destructive carbon emissions unleashed by northern countries, a development discourse originating in those same countries that blamed the agrarian poor in the Global South for environmental destruction, and a focus on outside intervention as the solution to degradation. To assert in response to the Anthropocene concept that Europe or capitalism is the responsible agent of environmental violence perhaps doesn't go far enough on this point: it fails to hold northern environmental NGOs to account for the adoption of neo-Malthusian development precepts that extended a colonial history of blaming poor and minoritized populations for degradation, which compounded neoliberal assaults on their livelihoods. This occurred even though northern states were the main beneficiaries of the so-called great acceleration of carbon-based development in the late twentieth century. Given the significance of this recent pre-history of the Anthropocene concept, it is necessary to think in multiple historical frames—in terms of the deep histories of capitalism and climate change as well as the recent history of neoliberal environmental thought—in order to recognize how different processes of both accumulation and knowledge production are distorted by universalist thinking about climate change.

If the long time span, social and ecological complexity, and widely distributed impacts of climate change highlighted by Anthropocene discourse teach us anything, it is that the circuitous routes of environmental violence produce persistent crises of representation, creating the need for multiplying our knowledges and narratives of how humans and other species interact through the carbon economy. This includes narratives that operate at a more intimate scale than planetary accounts of environmental or political-economic change and that force us to consider the local and transregional circuits of environmental harm that frame contemporary environmental justice struggles. Replacing the abstract universal of the Anthropocene with a large-scale notion of capitalist or colonial responsibility may help us unveil biases and distortions in the universal invocations of human responsibility for environmental destruction, but it is less useful in helping environmental justice movements (1) develop a detailed critique of how the fossil fuel economy—specifically oil—develops particular ecological, social, and geopolitical harms and inequalities, and (2) promote grassroots efforts to restore interspecies relationships and ways of living that have been displaced by settler colonial processes of accumulation, extractive land use, and industrialism.

In parallel, it is necessary to take more seriously the complex relationships that southern publics have to the carbon economy. One response to critiques that center the responsibility of the rich northern countries is articulated repeatedly by U.S. and European climate negotiators: how can there be serious international action against climate change without addressing states like China and India, whose total fossil fuel emissions are surpassing or predicted to surpass those of the United States? Such criticisms have been posed disingenuously by U.S. representatives in order to influence and ultimately dismantle Kyoto framework treaty negotiations, and it is important to remember that countries that have large export-oriented manufacturing bases use a proportion of their carbon emissions to meet the demand of consumers elsewhere—most notably the United States. So it becomes difficult to measure what proportion of Chinese, Bangladeshi, or Vietnamese carbon emissions are actually produced to sustain the massive U.S. or European consumer markets or to meet the growing demand in large consumer markets like those in the Persian Gulf, or in high-consumption global cities like Singapore. Nonetheless, the roles of states in the Global South as producers and consumers in the carbon economy must be taken seriously in order to account for the integration of formerly colonized nations into current circuits of capitalist development and energy use. It would be a mistake to avoid addressing how states and publics in the Global South experience constrained yet real forms of carbon-fueled agency despite the fact that the United States and Britain bear most historical responsibility for anthropogenic carbon emissions, and despite the fact that Australia, the United States, and Canada remain the highest per capita consumers of fossil fuels. Focusing attention on the ecocides wrought by European colonialism, slavery, and mercantile empire over centuries makes sense as a response to simplistic notions of universally shared environmental risk and responsibility under the banner of the Anthropocene. But it does not by itself give us the tools to imagine how to challenge capitalist forms of social reproduction that stubbornly tie people across differences of race, class, and nation—including large populations in the South—to destructive fossil-fueled economic practices.

Oil: "Corpse Juice" of Planet Earth

Planetary Specters suggests that thinking carefully about the geographical shifts enabled by the oil economy is key to understanding how environmental inequalities are reproduced through international economic arrangements. How can we make sense of a situation in the twenty-first century in which we witness unprecedented levels of transborder mobility and communication at a moment when so many of the world's poor and displaced people experience life as a shrinking horizon of habitation? This is not only a condition of people immediately affected by sudden weather disasters but also a broader problem of forces of infrastructural, housing-related, and ecological displacement and concentration worldwide. What explains the overlapping forces of what geographers used to call the "time-space compression" of globalization (cosmopolitan capacities allowing some people to cross borders and speed communication at ever-faster rates) and time-space expansion (containment of other people to shrinking horizons of movement, reproduction, and sociality) that attends carbon-fueled climate crisis?[42] The mythic invocations of neoliberal speed in globalization discourse have their cognate in climate discourses about the "great acceleration" in human development that generated the Anthropocene. Great acceleration discourse suggests that the acceleration in human development came with the price of amplified levels of carbon and other forms of widely distributed environmental waste which increased in magnitude in the second half of the twentieth century.[43] By any realistic measure, these forces of development are unequally shared, ensuring that agrarian populations most severely affected by climate change are also those least likely to have benefited from carbon-fueled regimes of capitalist accumulation. It is in this unequal share of the costs and benefits of neoliberalism, borne across the international division of labor, that we can begin to glimpse how neoliberal capitalism operates as a racialized economic system.

On this point, it is necessary to attend to how two particular nodes of the carbon economy—oil extraction and atmospheric carbon waste—configure migration processes. The twin mobilities of communications networking and carbon-fueled transit that allowed multinational corporations to transnationalize their manufacturing, labor, finance, and trade practices in the late decades of the twentieth century are at present largely dependent on coal- and oil-fueled infrastructures. These infrastructures enable rapid control of production and transoceanic shipping of goods along ever more complex transborder supply chains. Oil was particularly central in these transformations, as it allowed the proliferation of off-grid, transnationally mobile power sources that could help smooth transit and other logistics of highly decentralized production and trade.

Energizing growing scales of transborder trade even as its high price made it difficult for many countries to access, the political economy of oil led to enforced loans, which required poor countries to undertake policy transformations known as structural adjustment. In this process, pushed by U.S.-based conservative economists and dubbed "the Washington Consensus," loan conditions imposed on poor countries by international finance agencies like the International Monetary Fund and the World Bank required states to open up to cheap imports from abroad and to make massive cuts to government social services for health, education, and the environment. This process, which moved poor countries away from their postcolonial state-controlled markets and toward private capitalist control of larger and larger parts of social life, trapped many poor countries that lacked an industrial base in unpayable national debts and undercut the potential of their largely poor, agrarian, and rural populations to compete in the increasingly international markets that took shape domestically. The effects of this corporate globalization process on agrarian and working populations in many states in the Global South were devastating, displacing traditional industries and subsistence lifestyles due to importation of basic goods. Large numbers of working and poor people were increasingly rendered as reserve surplus labor.[44] As their countries were opened to greater volumes of international trade controlled by multinational corporations headquartered in the North, these populations were displaced from agrarian labor, which could no longer sustain livelihoods in the face of cheap agricultural imports, and were pressured to migrate to cities where they sought waged industrial work. Although the neoliberal revolution was uneven—it did not end complex strategies undertaken by postcolonial states attempting to navigate alliances and redirect decision-making authority and economic power to the South—forms of solidarity among African, Asian, and Latin American nations came under duress as the financial basis of U.S. empire encouraged competition for dollars and basic resources.[45] This neoliberal transformation from the late 1970s to around 2000—enabled by oil as both a form of energy and a finance circuit sustaining a relatively high price for the North's exported goods—caused large amounts of displacement. And this displacement was not mainly experienced as enhanced mobility in poor countries but as displacement from the rural agrarian base and concentration in urban peripheries, an alienation from land and kin. The resulting remittance economies—wherein displaced industrial workers become distance providers for rural family members who remain in the natal community—demonstrate that the demand to migrate is also often a profitable source of labor exploitation because it establishes dependencies on spatially alienated work for the reproduction of entire communities. All of this takes shape in a context in which migrants displaced to urban industrial zones are often viewed as racially, ethnically, or linguistically different, their labor configured as alternately exploitable and excludable as a condition of difference from those entitled to habitation, rights, and social welfare.

The extraction and waste effects of oil simultaneously produced large-scale forms of environmental destruction affecting racialized minorities and Indigenous nations during this period of neoliberal globalization. The violence orchestrated by Royal Dutch Shell in the Niger delta, including the repression of Ogoni activists and the assassination of Ken Saro-Wiwa, is one of the most violent episodes of extractive enclosure that arose during the expansion of oil-fueled neoliberalism.[46] But it was not the only one. Oil production is often attended by transformation of the landscape, including displacement of local people; pollution of land, air, and water; and greater risk of destruction through transport spills. Such extraction risks have only increased in the first two decades of the twenty-first century as the Washington Consensus came under criticism from labor, environmental, farmer, and Indigenous movements worldwide. During the past two decades, the production of shale oil, tar sands oil, deep sea oil, and other "unconventional" hydrocarbons took on an increasing share of the global energy market. This enabled historical energy importers like the United States and Canada to proclaim independence from Gulf oil in the wake of the 9/11 attacks; when Athabasca oil sands in Alberta were reclassified as reserves by Canada in 2002, the country was defined overnight as the world's second largest holder of oil reserves, a large proportion of its production destined for export to the United States. Meanwhile, the use of oil to fuel transnational shipping and growing small vehicle use was a key contributor to the increase in atmospheric carbon concentrations that exceeded four hundred parts per million in 2015. The widely publicized struggles of Indigenous nations in Canada and the United States against tar sands production and pipelines represent an attempt to connect the extractive infrastructure of oil to its networked effects in settler colonial environmental and economic systems.

Once categorized as an "unrenewable" resource, oil can now to an extent be "renewed" beyond earlier forecasts of available reserves. Deep sea drilling, which has been the prime conduit for the expansion of production in the United States (and also Brazil, Norway, and other countries), has produced a series of underwater oil spills, adding to the already significant risk of tanker spills as oil is shipped across oceans. These oceanic spills are responsible for destruction of coastal ecosystems, from the 1969 Union Oil spill in Santa Barbara, California, to the *Exxon Valdez* tanker spill off the Alaska coast in 1989, to British Petroleum's *Deepwater Horizon* disaster in the Gulf of Mexico in 2010. Beyond spills, risks of a globalized oil economy are increasingly reflected in the mining of bitumen in the tar sands, which is significantly more destructive to land, water, and air than conventional oil production. For much of the twentieth century, the politics of oil were configured around the potential of its declining supply, apparent in the Nixon administration's concern in the 1970s about the "fragile and finite" condition of "natural resources."[47] Although bitumen, as the most viscous layer of oil, has historically been used for tar production rather than for energy, oil companies discovered that it was possible to chemically transform it by taking over the geological processes that over millennia transform organic matter into crude oil. By using large amounts of water to filter the oil sands from the bitumen, oil companies are able to synthetically speed the process for producing oil, leaving behind large pools of wastewater that contain the toxic traces of the bitumen sand: nickel, vanadium, lead, chromium, mercury, and arsenic.[48]

By taking over the geological processes that produce hydrocarbons over the long span of geological time, unconventional oil and gas in the form of cracked shale oil, synthetic tar sands bitumen, and fracked natural gas represent an innovation on the temporal management that oil enables. Here is one way we might describe the energetic transformations enabled by conventional oil production: in layers of sedimentary rock, fossils of dead animals, plants, and microbial life are compressed and heated over millennia, transformed from solid to liquid deposits, the viscosity of which allows oil to seep across rock layers and pool in areas that may or may not lie close to the lithosphere, where humans can drill into them. Once discovered, collected, and refined, this sludgy carbon material is then distributed widely and combusted to power a number of high-energy-input human activities, from automobile and airplane transit to electricity generation for industrial machinery and heating. During this process, there is a transfer from the development of carbon capacities over the slow geological time of compression and heating of dead biological matter to the rapid acceleration of transportation and communication enabled by combustion, a time travel from the extinct bodies percolating in the deep time of geology to the viscous combustion of the phrenetic time of globalization's time-space compression.

The innovations of cracked and fracked hydrocarbons, as well as of deep sea drilling, add complexity to this temporal process of conventional oil, creating a secondary feedback loop in the system. Drawing from the profits of conventional oil, oil firms transform the upper layers of the lithosphere into a kind of machine for filtering, compressing, and heating partially composed fossil fuels into synthetic carbon in shale oil and bitumen processing.[49] By using highly carbon-intensive processes to generate new capacities for extraction, the time travel of long extinct bodies to contemporary fossil-fueled mobilities in turn shrinks the deep time of geology toward the industrial time of the global and allows a momentary capability to "renew" the nonrenewable resource: to exert the force of fossil-fueled racial capitalism back into the deep structures of the earth. This circuit seeks to bypass the geological limits to capitalist speed, displacing the supply crisis of peak oil and stretching the time of widely dispersible neoliberal supply chains toward an indefinite horizon. Meanwhile, the feedback loop that allows oil to reproduce itself, recruiting more and more bodies of the dead—human and nonhuman—into the enhancement of neoliberal mobilities in the present, massifies oil's atmospheric waste: emissions continue to rapidly rise in concert with economic activity generally. There can be no transnational trade in its current form without it, and economic growth for any country today necessarily correlates with the more rapid and devastating effects of climate change.

These aspects of oil's widely distributed infrastructural significance; its environmental feedback loops; its subsurface deposition; and its background capacities for enabling varieties of human representation, transit, and accumulation all help explain why energy humanities scholars who research petrocultures describe oil as a confounding material underpinning contemporary life. For Imre Szeman, the oil question is not an ontological or a historical question but an "epistemic" one, for human societies are "so saturated with the substance that we cannot imagine doing without it." As such, oil forms "part of our knowing," a material reality that shapes our representational lives and elaborates key categories of liberal interconnection, including freedom, mobility, growth, and futurity.[50] For writers and artists including Stephanie LeMenager, Edward Burtynsky, and Amitav Ghosh, oil forms such a widely distributed part of the material conditions of culture but recedes into the energetic background in ways that make it difficult to represent.[51] Such an argument parallels how environmental humanities scholars write about climate change as a crisis of representation: climate change seems to challenge the scales at which humans normally conceive of time and space, operating at the geological scale of Earth. And yet oil has been at the center of formal political struggle for a century, ranging from colonial prospecting to environmental justice campaigns against drilling to conflicts over access to underground stores among oil-producing states. Oil animates national energy economies and transnational commodity and labor chains, even as it is rendered in generic units for international trade and environmental surveillance.

Although it conventionally avoids focus on the extraction side of ecological violence and displacement, climate migration discourse focuses attention on widely distributed effects of climate change—the waste products of the extractive carbon economy. And in terms of aggregated effects on life systems, the impacts are being felt across the planet. Atmospheric carbon emissions are increasing the number of deaths due to particulate pollution worldwide, with an estimated 6.5 million premature deaths in 2015.[52] Extreme heat events generate crop failure, with the likelihood that current rates of warming will reduce crop yields in Central America and the Caribbean by 50 percent by 2050. Freshwater availability and fish stocks are already on the decline due to ocean heating, loss of glacier volume, and desertification.[53] The violent effects of such forces are undeniable, yet the ways in which they affect migration requires assessment of their correlation with other capacities for habitation, including infrastructure, industry, employment, and economic resources. But given the manner by which the geology of oil energizes such circuits of atmospheric waste toward existing extractive forms of inequality, violence, and disposability, Reza Negarestani characterizes oil as the "corpse juice" of capitalism, a material that connects neoliberalism's cosmopolitan capacities for mobility and exchange to the death-dealing effects of war and ecological collapse.[54]

Planetary Specters

When considering the recent proliferation of climate refugee discourse, it is necessary to keep in mind that new rights of asylum emerge out of conflicted contexts of historical change. One element of climate migration discourse is a fear of blurred borders and rising "extremism" coincident with the failure of states to control population movement and implement successful development. On the other hand, a vision of climate migrants as fleeing shrinking zones of habitation configures a form of humanitarianism that operates to salvage peoples and lifeworlds viewed as inevitably dying, as doomed to the ravages of a carbon economy that cannot be brought under control. As such, climate migration discourse participates in broader discourses of Indigenous salvage that piggyback on a history of anthropological depictions of the agrarian poor's vulnerability to environmental degradation. At times, these contradictory depictions of swarming threats and romantic depopulation converge in representations of the very same racialized groups, as in the representations of coastal farmers in Bangladesh explored in chapter 3.

To the extent that the migrating body can provide a racialized focal point to distill a variety of crises that reflect the destructive inequalities of neoliberalism, its incorporation into the law signals a fantasy of redress that seeks to redraw borders by offering climate-affected peoples the supposed gift of humanization by states that promulgate border imperialism. As such, the spectacle of inclusion that constitutes the ethical demand of climate migration discourse does not disrupt what Christian Parenti calls the "politics of the armed lifeboat"—policies aimed at creating privileged zones of safety against an outside world conceived as ever more chaotic, ungovernable, and unable to consistently support life.[55] The metaphor of the armed lifeboat insists further that there is a relation between the safety of the boat and the violence outside. Through its militarized borders, it invests in a defensive posture that requires maintaining ever stricter separations between the securitized inside and the abandoned outside.

Such forms of racial securitization, evident in political contests over both climate change and immigration, demonstrate how the myths of humanitarian redress promulgated by the wealthy countries advance a geopolitics that reproduces neocolonial international divides; as a liberal project that forestalls the broader liberatory promises of decolonization, human rights law according to Randall Williams maintains violent neocolonial divisions of power in the world system.[56] Jamaican philosopher and dramatist Sylvia Wynter argues further that twenty-first-century "struggles over the environment, global warming, severe climate change, the sharply unequal distribution of the earth's resources . . . are all differing facets" of an "ethnoclass" struggle that lies at the heart of the colonial politics of the human, a struggle most spectacularly witnessed by the "new poor" who are defined "at the global level by refugee/economic migrants stranded outside the gates of rich countries."[57] Wynter configures climate change and the expansion of the new global poor as interrelated effects of colonial modes of thought and politics that separate human from nature, establish a racial geography of habitable ecological zones, and configure the African diaspora as the lowest rung of humanity against which northern states attempt to recruit migrants in reterritorializing hierarchies of the racial state. Thus viewing contemporary environmental crises as effects of a planetary racial and class order, Wynter challenges us to develop new accounts of human social and ecological relations that question assumptions of capitalist technological development and freedom.

The forms of violence produced by oil and other fossil fuels do not go without ideological and political contestation, as shown by the small island activists in Kiribati, the Oceti Sakowin and Wet'suwet'en pipeline resisters in North America, and other Indigenous environmental activists worldwide. Environmental justice organizations have focused significant effort on researching and transforming industrial practices that concentrate pollutants from petroleum production and other industrial production in minority communities. Such efforts are complemented by antiwar activists who have criticized the U.S. wars in the Persian Gulf region over the past two decades as cynical ploys using the 9/11 attacks as justification to control the geopolitics of the world's largest oil-producing region. If public concerns are growing that climate-driven migration might produce instability or conflict, struggles over the control of fossil fuels have also helped generate discourse on the relationship between Global North and Global South and on the links between migration and oil. Oil's role in forms of capitalist development and U.S. empire creates complex displacements and political struggles that have been key touchstones in debates over models of social transformation on the Left. While such debates have at times exhibited some of the limitations in activists' ability to develop adequate accounts of the relationship among race, oil, and migration (philosopher Michel Foucault's late writings are a case in point),[58] Indigenous- and minority-led campaigns such as Idle No More signal that a new generation of activists are emboldened to challenge the global infrastructure that smooths the reproduction of oil-fueled racial capitalism.

In solidarity with such struggles, *Planetary Specters* develops an analysis of climate migration discourse with an eye to how changing configurations of political economy, geopolitics, and warfare helped to create contemporary pathways for environmentally affected migrants. Focusing on the ways that the rise of neoliberal forms of capitalism across Asia—including oil-based finance and militarization in the Persian Gulf, labor diasporas emerging in South and Southeast Asia, and growth in East Asian manufacturing—are central to understanding how environmental forces affect migration, the book argues that oil has provided a new basis for racial capitalism's ability to recruit migrants as exploitable labor over the past half century, both as the nature of U.S. empire has shifted and as environmental processes have intensified pressures on the agrarian poor.

Chapter 1 argues that the figure of the climate refugee has become one of the central concerns of interdisciplinary climate research; this figure indexes shifts from mitigation to adaptation and from multilateral governance to state-centered security discourse in climate policy. Recounting the genesis of the figure of the environmental refugee in development studies and NGO research on environmental degradation, I suggest that a narrow state-centered view of climate migration as an international security crisis produces racialized geopolitical divisions, border militarization, and Malthusian precepts about rural overpopulation in the South. With the configuration of new security discourses around climate migration, the specter of Islam and the speculation of mass migration reconfigure how race is conceptualized through discourses that aggregate potential environmental and social differences arising from extreme weather events. Such trends are transforming some arenas of environmental thinking and activism—trends that environmental justice research should resist in order to develop a broader cartography of carbon's human and ecosystemic costs.

Chapter 2 argues that theories of racial capitalism offer a productive method for moving beyond stereotyped and securitizing imageries of climate migration. Revisiting classic works on racial capitalism, including Cedric Robinson's *Black Marxism*, the chapter analyzes race's connection to the financial, geologic, and atmospheric formations of the carbon economy, with a special focus on West and South Asia. A theory of racial capitalism extended to considering how oil infrastructure generated neoliberal climate crisis allows integrated analysis of carbon-based neoliberal trade, the development of the South Asia–Persian Gulf migration corridor, and processes of atmospheric waste amplification that cause desertification, sea level rise, and intensified weather disasters. These processes intensify migration-driven forms of racial differentiation, as populations facing the violences of extraction, land displacement, labor exploitation, pollution concentration, crop failure, flooding, and debt experience the unequal effects of accumulation and unequal access to forms of state redress. The chapter concludes by examining how such racialized climate inequalities influence the recent emergence of concepts of natural capital, which attempt to extend capitalist forms of value to nature as a means of reconceptualizing wealth.

Chapter 3 focuses on how oil-driven development laid the groundwork for Bangladesh to become the world's first "climate migration" hot spot. Critiquing visions of climate-driven mass migration, ethnic conflict, and Islamist terrorism narrated by Bangladeshi government officials, international security experts, development NGOs, and journalists, the chapter documents how the Bangladeshi state incorporated outmigration to the Gulf into its plans for neoliberal development beginning in the 1970s, which shaped the migration routes that are increasingly traveled by today's flood-affected migrants. The chapter argues that weather-induced migration is currently moving along the same routes and networks created by the oil boom and the rise of neoliberal trade, with the largest flows moving internally from rural areas to the two major cities of Dhaka and Chittagong, and external flows heading to the Gulf. These dynamics challenge conventional thinking that ambivalently regards climate refugees from Bangladesh as either tragic victims of Western overconsumption or romantic figures of adaptation. Tracking how such representations of the climate refugee in Bangladesh are being mobilized by the state and NGOs to intensify racial capitalism's processes of agrarian displacement, the chapter demonstrates that emergent climate adaptation schemes are configuring migrants as sites of human capital that can be cultivated by NGOs and southern states.

Chapter 4 examines how the ongoing civil war in Syria has been configured by security experts, climate scientists, and Western journalists as a "climate war," particularly in the wake of a 2015 study claiming that drought was the trigger event for the Syrian uprisings.[59] Detailing how narratives of racial disability are used to buttress claims that the Syrian uprisings were in essence climate conflicts requiring intervention, I argue that visions of climate-driven resource conflict attempt to skirt the complex political and economic underpinnings of Syrian resistance to the Assad government. By focusing on accounts of aerial bombing and infrastructure collapse during the war, it is possible to give an alternative account of environmental factors related to the Syrian and Arab uprisings and to witness other histories of environmental resistance in the conflict, including efforts to redistribute land and develop ecological collectives in Rojava. This reflects some potential alternatives to narratives of scarcity-driven climate migration, as forms of ecological thought can envision collective approaches to land, wealth, and interspecies relations.

Although it remains to be seen whether more changes in weather systems will establish new migration practices particular to the era of rapid climate change, this book argues that human movement builds on established networks of affiliation and knowledge about mobility. This makes it less likely that climate change will produce the widespread social breakdown envisioned by security analysts and right-wing, xenophobic critics of climate change. Resisting the racial stereotypes that are visited on climate migrants in such discourses, one of the lessons of weather disasters is that mutual aid and cooperation remain key outcomes when climate change challenges state-run infrastructure networks. This means that analysis of migration in climate-affected locations can become a key starting point for alternatives to the current carbon-fueled order of racial capitalism. The conclusion addresses this potential by analyzing recent migrant narratives that attempt to foreground the integration of geopolitical and environmental concerns. In so doing, these narratives help us see some pathways forward for critical studies of race, migration, and security—pathways that move away from statist forms of racialized crisis management and invoke practices of collectivity that transcend the armed lifeboat. In the process, they configure forms of habitation against the notion of a permanently sinking future.

CHAPTER ONE

Race, Insecurity, and the Invention of the Climate Migrant

"Mobility is emerging as the human face of climate change," write the authors of the 2018 World Bank Group report *Groundswell*, which predicts that the rise in average global temperatures will displace as many as 143 million people in Africa, South Asia, and Latin America by 2050.[1] Building on research about environment-migration connections by climate activists and scholars, today international institutions (the World Bank, the International Monetary Fund, and the United Nations) and U.S. and European security agencies publicly discuss mass environmental displacement as a potential security crisis producing instability and conflict. Such public discussions have been accompanied by a series of court cases that explore whether there may be a basis for asylum among environmentally displaced people. Such pronouncements of planetary crisis help to propel interest in climate-driven migration across an increasing range of media as well as policy, legal, and academic writing. At the time this book was written, atmospheric carbon had reached a concentration of 409 parts per million—a three-million-year peak, exceeding levels that scientists have discussed as thresholds for rapid global warming and destructive sea level rise. The massive scales and rapid ecological shifts of climate change and related human-induced ecological processes are now being recorded in all regions of the planet, transforming areas of settlement in locations most vulnerable to storms, flooding, desertification, and other disasters. This undoubtedly makes displacement and migration two key areas in which climate change is likely to exacerbate social inequalities affecting poor, displaced, and minoritized populations.

Yet as the scale of climate change disasters propels public debates over the protection of ecosystems, species, and human communities, it is necessary to recall that the forms of social change that emerged with the fossil energy systems of coal and oil, and that were embedded in routes of globalizing trade and extraction, have long produced massive migration flows. Whether from rural to urban areas internal to nation-states or across national borders, migration—especially among people dependent on agriculture for survival—has long been one symptom of the unequal forms of human development. Such inequalities have marked the various phases of capitalist expansion since the sixteenth century but have emerged with particular planetary force during what environmental historians have recently named the "great acceleration" in human-induced ecological changes since World War II, during which industrial output, resource depletion, and carbon emissions have grown rapidly.[2] From this vantage, it appears that the history of colonial energy use, which intimately entangled European and U.S. projects of industrialization and state formation with labor, resources, and social movements from the Global South, is a necessary backstory for understanding contemporary imaginaries of environmental crisis.[3]

Is there something particular about carbon-fueled climate change in the twenty-first century that makes the experience or effects of migration different from earlier eras? Why is mass migration increasingly understood as an inevitable risk of climate change and, in some contexts, one of the necessary adaptation strategies to it? And if climate change is, as security analysts and economic think tanks increasingly warn, propelling an increase in global migration, how can we understand how this current moment of disaster migration emerged out of larger systems of human settlement, state formation, energy use, and exchange? Noting the tendency of some journalists and security experts to depict climate change as a new and massive force propelling migration, this chapter tracks the invention of the figure of the climate migrant in environmental policy and journalism discourses of the last three decades. It argues that the figure of the climate migrant has emerged as a media icon not because the nature of the weather or of ecological destruction has fundamentally changed the nature of human mobility but because it offers a rhetoric about migration that appeals to a set of racialized presumptions about human conflict and population dynamics that fit into increasingly apocalyptic and conservative northern political imaginaries of climate change's destructive social effects. By positioning displaced people in the Global South as disabled victims of geophysical processes rather than subjects responding to political and economic inequalities, climate migration narratives that have emerged in the publications of think tanks, development researchers, security agencies, and journalists often fall in line with neoconservative forms of risk prediction, securitization, and geopolitical mapping, which have become increasingly prominent state logics. Although these trends are most evident in the United States, Europe, and other wealthy countries, climate security discourses increasingly infuse journalism and policy discussions in the Global South as well. With little traction in courts for elaborating new rights for so-called climate migrants, recent climate migration discourses have been immersed in a new security thinking, which is oriented, on the one hand, by highly racialized speculations of Islamist militancy and, on the other, by posthuman social mapping techniques that transform the ways that vulnerable groups are conceptualized in emergent forms of surveillance and governance. From this vantage, it is possible to see how the turn toward security thinking in environmental politics recapitulates a longer history of racial governance that configures environmental degradation and population growth in the South as the causes of underdevelopment and conflict despite the fact that U.S. and European powers have been deeply involved in creating mass migration over the past half century.

By mapping this broader context, I reflect on how narratives of twenty-first-century climate migration as an exceptional and sudden form of migration build a racialized framing of environmental disaster. This framing obscures public understanding of the imperial force of carbon in the making of the current geopolitics and racial labor structures of transnational capitalism. Although later chapters will focus most closely on Syria and Bangladesh—two locations in which discourses of climate migration have had particular influence in international media—the history of the racialized discourse of climate migration (detailed in this chapter) as well as the colonial roots of oil's international inequalities (detailed in chapter 2) have broad relevance in all places in which climate change affects livelihoods and habitation. Reckoning with the ways that climate migration discourse emerges out of longer histories of humanitarian and environmental politics, as well as the ways it is framed by emerging discourses of security, is necessary in order to formulate challenges to the current destructive order of carbon-fueled social reproduction and the inequalities it generates.

Figuring the Scale of Climate Insecurity

Today, the figure of the climate migrant or climate refugee has become one of the most visible public icons of climate change as an environmental injustice. Whether located among the growing masses of people migrating north from the Global South in Latin America, Africa, or Asia, or among vulnerable coastal populations at risk of internal displacement, the climate migrant appears amidst growing attention among journalists and international institutions to major weather disasters, as images of human settlements devastated by cyclones, floods, and rising seas circulate widely in digital media. Whether reporting on catastrophic infrastructure breakdown in Puerto Rico, hurricane flooding in Cuba, river flooding in Bangladesh, drought and crop failure in Syria and Somalia, sea level rise affecting coastal cities and small islands, or massive glacial and mountaintop melting in the Himalayas, media attention to the destructive impacts of climate change increasingly signals fear of infrastructure breakdown, famine, and the potential for internal and cross-border displacement.

As a matter of research, policy analysts have worked at length to quantify the scale of potential displacements. As climate migration discourse has moved from a niche arena of environmental policy analysis to broader discussions of economic and security policy, increasingly dire pictures of the global situation have been forecast. In the past decade, international policy analysts in the fields of development economics and international relations have produced a steady stream of reports on the displacement crisis generated by weather disasters and intensified by rising global temperatures. The Internal Displacement Monitoring Centre (IDMC), a key think tank based in Geneva that produces research on climate-driven migration, suggests higher degrees of human mobility generated by climate than does the World Bank's *Groundswell* report. In 2018, the IDMC reported over 12.4 million new internal weather-disaster-related displacements in just the five countries with the largest displaced populations: the Philippines, China, India, the United States, and Indonesia.[4] Set amidst a much larger world picture of mass migration, in which over sixty-five million cross-border migrants and many more internally displaced persons are on the move, images and narratives of climate-driven migration increasingly weave together evidence of vulnerability and desperation among the world's poor with claims that displacement is caused by the widely dispersed effects of increased atmospheric carbon.

If much of the effort by climate scientists and policy experts to generate interest in climate change in the 1990s and 2000s focused on developing a public discourse around human responsibility for the destructive effects of carbon emissions, by the current decade the figure of the climate migrant or climate refugee has coalesced as a primary symbol of climate change's threat to human life and social stability, particularly among the poor in the Global South. In the process, the icon of the climate migrant has moved from an obscure figure in development studies discourse to a primary object of state and NGO concern. When the Egyptian development scholar Essam el-Hinnawi first coined the term "environmental refugee" in a 1985 policy paper published by the United Nations Environment Programme, the problem of environmental migration emerged out of debates within development studies. At that time, UN development experts sought to prevent the world's agrarian poor from degrading their resource bases, which, it was thought, would exacerbate development problems that made them subject to displacement: "Prevention is urgent, as the frequency and severity of disasters are increasing to the point of being unimaginable. In large parts of Asia, Africa, and Latin America, the ecological base for human existence is being damaged so badly that it can no longer support its growing populations. Most disaster problems in developing countries are unsolved development problems. . . . The question of averting natural hazards and halting the flood of environmental refugees is, therefore, a question of environmental management."[5] In this passage, el-Hinnawi configures problems affecting human capacity for settlement—poverty, infrastructure challenges, and environmental trigger events—as functions of the society-environment relation that express the ability of a natural "habitat" for humans to provide capacity for social reproduction. As such, el-Hinnawi follows trends in development and environmental literatures that express population as a determining factor of potential adaptation to environmental change.

From the early writings on environmental migration, questions of scale and widespread development problems were central to the discourse. Since such concerns were ripe for thinking in terms of crises that would cross borders, it is perhaps no surprise that environmental migration reporting moved in the direction of security thinking, even before the 9/11 attacks made this a central preoccupation of states. After el-Hinnawi, the British climate activist and scholar Norman Myers is often credited as the first author to put climate-induced migration on the international policy agenda. Myers's series of papers and reports on the topic is notable in that the ostensible humanitarian focus on climate refugees beginning in the 1990s is supplanted by an emphasis on security by the 2000s. This change in emphasis indicates how liberal environmentalist thinking had at its roots assumptions about how population pressures on limited environmental resources could produce degraded land and cause resource scarcity resulting in social conflict. As such, climate migration thinking has always been a ripe arena for the development of securitizing discourses and technologies.[6]

Although discussion of the rights of migrants affected by climate change is ostensibly a liberal political project, the rise of climate migration discourse coincided with the end of binding emission reductions in the Kyoto framework negotiations and the concomitant rise of climate security discourse, especially in the United States and Europe. Climate migration discourse—like all expert discourse on climate change—is not simply produced and disseminated according to scientific rationales. Conflicts between states as well as negotiation among development, conservation, and carbon mitigation goals influence the rhetorical strategies used by state officials, NGOs, and scientific experts to speculate on climate change impacts. Since at least the 1992 United Nations Earth Summit, which invoked visions of a more sustainable, environmentally friendly capitalism as a solution for merging the UN's conflicting conservation and development goals, governments in the Global South have had a role in articulating demands that poor countries be protected from short-term economic losses of reducing fossil fuel consumption and deforestation. Yet by the early 2000s, much of the international discourse on climate was dominated by large polluters, especially the United States, the European Union, China, and India. The role of U.S. negotiators in minimizing U.S. responsibility for carbon mitigation in Kyoto Protocol negotiations led to an emphasis on security discourses that appeared more palatable to unilateral defense agendas. Rather than mitigating carbon emissions, such schemes focused on adapting to rising global temperatures and establishing defenses against some of the major impacts of climate change that were increasingly viewed as inevitable.

Meanwhile, the Far Right has openly embraced discussion of climate adaptation precisely because of fears of "race suicide" and the potential for climate-driven migration to affect the demographics of U.S., Canadian, and Australian settler colonies and European states. To the extent that climate migration discourse increasingly makes the argument for environmental solutions on the traditional grounds of the Right—security, policing, military intervention—it has recently been taken up by the Far Right as one thread of a broader anti-immigrant discourse. The white nationalist journal *American Renaissance* recently published an alarmist article suggesting that climate change was contributing to so-called demographic threats against majority-white states.[7] Mette Frederickson of Denmark's far right People's Party notes, in the context of the recent refugee streams to Europe, that "climate change will force more people to relocate," a claim she combines with a warning that "the population of Africa is expected to double by about 2050."[8] Although such xenophobic arguments are emerging slowly out of parties that have historically been committed to denial of climate science, environmentalists in Europe are now noting that younger generations of activists are pushing "green nationalism" among far right, nationalist, and white supremacist political formations.[9]

Race, Gender, and Displacement

This xenophobic interest in environmental migration on the Right—anticipated by what el-Hinnawi's UN report envisioned as the possible "flood" of environmental refugees—is one reason that we must attend to how the climate migrant as a media figure may obfuscate the complex relationships among race, capitalism, and ecological precarity. As a figure, the climate migrant stands as an icon of social vulnerability, tied deeply to public images of racial differentiation. Yet the racial and gendered spectacle of the climate migrant can mask the structural and historical forces that have produced systemic vulnerability for a variety of subaltern national, racial, and class groups, especially with the rise of the oil economy in the twentieth century. Using stories of personal tragedy linked to images of dark-skinned migrants—mainly women and children from the Global South—struggling to maintain adequate shelter and food in settings of ecological crisis, climate migration discourse works to homogenize different situations of social conflict and vulnerability into a narrative in which geological and atmospheric processes are positioned as the true roots of displacement.

Take, as an example, the National Geographic Society's encyclopedia entry for "climate refugees" in its online resource library, published in March 2019.[10] The entry, aimed at primary school students, fails to actually define the term, opting instead to explain that climate change affects different climatic regions. But it does include an image of an unidentified veiled woman, who serves as an emblem of the resulting displacements. Centering her face, the photographer focuses on the furrowed brow and sharp gaze of the woman, whose dark skin is illuminated in the bright sun, the desertscape behind her blurred. The caption reads, "a severe drought forced this woman and her clan to move over 240 kilometers (150 miles) in order to find water." The entry does not provide a name or any details of the location of the image. But the reference to the "clan," to desert and drought, and to long migration routes in search of water echoes much earlier writing that focuses on the purported environmental causes of conflict in the Sahara and the Sahel regions of northern Africa. For decades, colonial anthropology and development studies literatures created romantic and exoticizing narratives that focused on the disappearing pastoral practices of African herders and other nomadic peoples.[11] Such narratives move between romantic visions of the tragic decline of African pastoralism to images of environmental and social degradation among rural peoples, marked by scarcity and social breakdown. As such, the National Geographic Society entry for

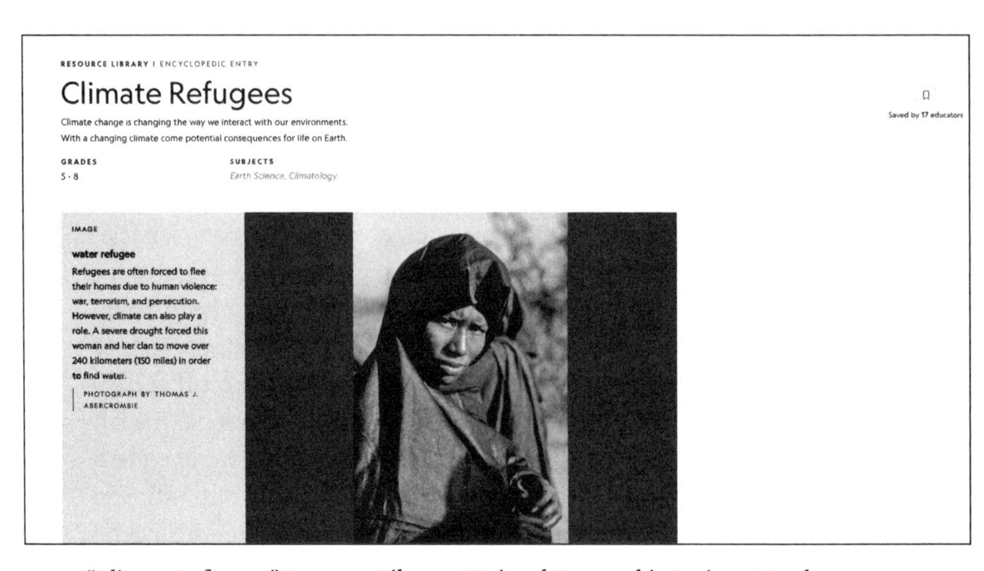

"Climate Refugees," Resource Library, National Geographic Society, March 28, 2019, www.nationalgeographic.org/encyclopedia/climate-refugees. Photo by Thomas J. Abercrombie.

"climate refugees" indexes the long-standing anti-Black racism of environmental degradation discourse, which is one source for contemporary climate migration narratives.

The fact that this "human face" appears without a name or a voice indicates the manner in which contemporary public representations of environmental change continue to build on generic tropes of colonial difference that render international crisis as gendered, raced, and classed, often located in rural geographic peripheries. The tragic images of farmers and other agrarian workers attempting to rebuild or maintain livelihoods following catastrophic weather disasters serve as potent symbols of planetary environmental destruction, as risks of social breakdown, and as icons for white rescue narratives invoking the vulnerability of women and children from the Global South.[12] This is not only true of publications like *National Geographic*, which have long been criticized for racist depictions of Africa and of Indigenous peoples worldwide, but also of global think tanks and NGOs that attempt to focus on climate-driven migration in order to make a case for either launching international efforts to mitigate climate change impacts on the world's poor or lowering barriers to migration in the Global North. IDMC's report outlining a case for new research on "slow-onset" displacements is a case in point. Displaying on its cover the 2017 image of a

Bina Desai, Justin Ginnetti, and Chloe Sydney, *No Matter of Choice: Displacement in a Changing Climate, Research Agenda and Call for Partners* (Geneva: IDMC, December 2018), 1.

woman photographed from behind—adorned with a bright yellow scarf, a green and white dress, and a bundle of sticks slung on her back—the blurred beige background displays a scene of drought that threatens habitation and life itself. The caption reads, "An internally displaced woman walked four hours with a heavy load to the village of Shisha in Somalia, crossing the mountains bordering the dry valley. From her 200 sheep, nothing is left. After building her tent, he [*sic*] intends to repeat the journey, bringing her children to the village."[13] In such narratives, environmental roots of migration are isolated from the complex networks of social, political, economic, and ecological relations that drive human mobility, in the process forging the icon of an economically and physically vulnerable dark-skinned third world woman as the embodiment of climate change's human impacts.

Parsing Environmental Causes of Migration

Images of individual migrants affected by environmental disasters often intend to communicate something of the human cost of climate change and to spur individuals and states into action to mitigate its worst impacts. So why should we be concerned about the racialized iconography of climate representation in media? The characterization of climate refugees institutes a form of racialization common to migration discourse more broadly, figuring migrants from the Global South as both vulnerable bodies and potential beneficiaries of humanitarian immigration policy within comparatively wealthy receiving states. Of course, the vast majority of these migrants never make it to the North; nations in the Global South become the host countries for four out of five refugees.[14] The operation of racial differentiation in migration discourse thus rehearses a fictive colonial division of geographic space that is crucial for maintaining nationalist myths in northern countries about the purported humanitarianism of liberal law, despite the histories of militarism, environmental destruction, economic exploitation, and political interventionism in which the northern countries have persistently helped to generate migration. Critical refugee studies scholarship—including a growing body of work focusing on southeast Asian refugee flows—highlights how the dual concepts of refuge (indicating the sanctuary promised by the receiving state) and refugee (indicating the displaced subject who purportedly appeals for humanitarian redress from the receiving state) co-construct an imperial discourse that reifies settler and imperial logics of humanitarian rescue. Such discourse tends to mask the histories of militarized violence that produce refugees as a racialized supplement to the political orders of empire. In *Body Counts*, Yen Lê Espiritu argues that "the figure of the Vietnamese refugee . . . has been key to the (re)cuperation of American identities and the shoring up of US militarism in the post-Vietnam War era. . . . Having been deployed to 'rescue' the Vietnam War for Americans, Vietnamese refugees thus constitute a *solution*, rather than a *problem*, for the United States."[15] Mimi Thi Nguyen describes this solution as a type of national myth-making in the form of "the gift of freedom" purportedly conferred on Vietnamese refugees by the United States; such a narrative of state care "discloses for us liberalism's innovations of empire, the frisson of freedom and violence that decisively collude for same purposes . . . because it may obscure those other powers that, through its giving, conceive and shape life."[16]

Yet the purported gift of freedom that receiving states may offer to the figure of the climate migrant becomes more difficult to define when one inquires into how such a migrant might be characterized as a legal subject. One clue to the difficulties of the construction of the climate migrant lies in the term itself. Under what circumstances would a person's mobility be labeled "climate migration"? Because of the complexity of migration decisions and the difficulty of isolating ecological influences on migration from others, the United Nations, the Intergovernmental Panel on Climate Change (IPCC), and other international institutions have not established accepted legal definitions of climate migration or climate refugees. Should all individuals who experience temporary or permanent loss of a home due to an extreme weather event in this era of accelerated climate change be designated as climate migrants? Such a definition would encounter the criticism that floods, hurricanes, and other weather disasters had been displacing human communities long before the era of anthropogenic climate change. As such, the IDMC uses "conflict" and "disaster" as the categories to identify root causes of internal displacement, with "weather-related disasters" (including extreme heat events, flooding, and drought) identified as the primary forms of disaster that produce displacement. According to the IDMC, 16.1 million out of 18.8 million people categorized as internally displaced in 2017 were affected by either floods or storms.[17] IDMC's work has been widely used to describe the growing threat of climate migration. Its annual Global Report on Internal Displacement website (https://www.internal-displacement.org/global-report/grid2020/) includes interactive maps displaying circles representing migration magnitude. These circles in most cases display a large blue section indicating disaster displacements, with a smaller section indicating conflict displacements. Data on disaster displacement of the sort visualized by the IDMC in its displacement map appear to suggest that weather-related displacement has outstripped displacement due to political factors, despite the fact that in the text of its reports, the IDMC gives contextual descriptions of how policies and social inequalities—including "poorly conceived urban planning" that affected people in flood-prone subsidized housing during Hurricane Harvey in Texas—affect the magnitude of displacement figures.[18] Although the IDMC does not use the terminology of "climate migration" to describe displacements from what it categorizes as weather-related disasters, its figures or similar ones reported by the United Nations are at times subsequently reported as such.

The World Bank's *Groundswell* report is also careful to distinguish environmentally induced migration from migration that can be directly linked to climate change (climate migration). But the need to parse differences between weather and climate in this distinction is also complicated by the need to address a number of other complexities in the categorization of migrants. The report includes a schematic representing the forms of environmental migration, emphasizing the gray areas between environmental displacement and migration as well as the interplay between the experience of mobility (as either forced displacement or voluntary adaptive migration) and

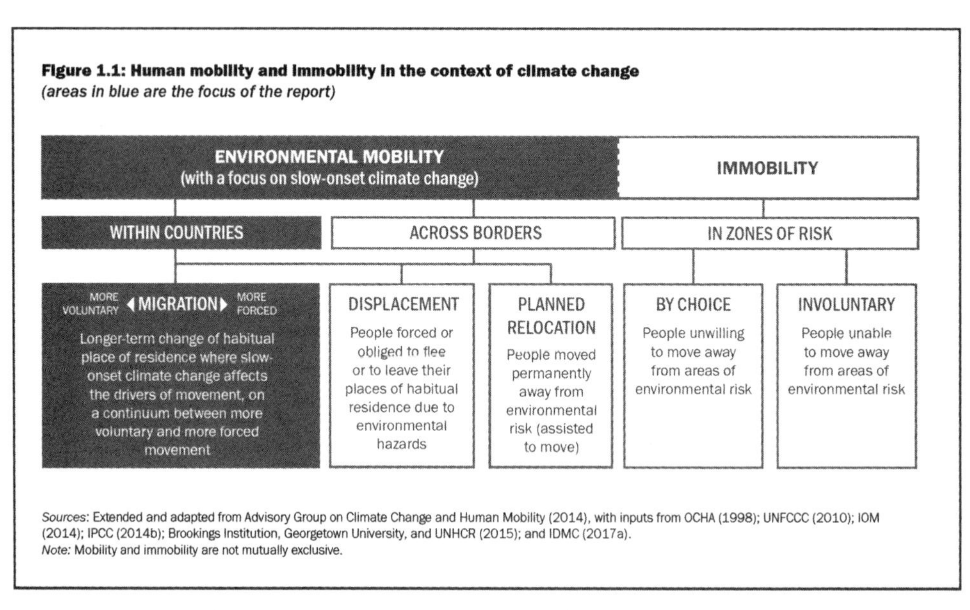

"Human Mobility and Immobility in the Context of Climate Change." In Kanta Kumari Ringaud, Alex de Sherbinin, Bryan Jones, Jonas Bergmann, Viviane Clement, Kayly Ober, Jacob Schewe, et al., *Groundswell: Preparing for Internal Climate Migration* (Washington, DC: World Bank, 2018), 4. © World Bank. License: Creative Commons Attribution (CC BY 3.0 IGO).

that of immobility (which is described as a stage of the experience of displacement). The chart is organized around three columns representing the momentum of people affected by environmental processes, categorizing migrants based on mobility within borders, mobility across borders, and immobility. In turn, each column depicts a binary between voluntary and involuntary movement or stasis. When do environmental inducements to migration result in "voluntary" migration, and at what point do such inducements become so burdensome that migration is "forced"? The fact that the chart is organized around the dialectic between choice and coercion, and is framed by the thresholds crossed when migration becomes transnational, signals that the concept of environmental migration reflects a crisis for liberal ideas of free agency, citizenship, and property in the self. As Malini Ranganathan notes, such liberal concepts of property and agency have been significant in constraining the potential radical force of environmental justice discourse as they conceive of environmental racism as either individualized harm or incidental externalizations of market processes.[19] Simultaneously, the particular scales and movements of environmental harm make it difficult to locate environmental racism in the normative time frames or human agencies of liberal politics. The subject of migration or stasis is potentially compromised by ecological forces beyond the subject's control, and yet the potential slow onset of such forces may make environmental determinants of movement imperceptible. Such attempts to parse the condition of the climate migrant reveal the embedding of the migrant in broader biosocial and geophysical forces inextricable from the mythos of individuated human agency.

Such problems of categorization reflect how the development discourse on climate migration recapitulates long-standing attempts by states to differentiate voluntary and involuntary migration as a political distinction. This issue is one currently faced by coastal Indigenous communities around the world. For example, the Biloxi-Chitimacha-Choctaw Indians of Isle de Jean Charles, Louisiana, are involved in a legal struggle over the U.S. federal government's relocation of their community as the island recedes into the Gulf of Mexico with the rising seas. They have been labeled in numerous media outlets as the first U.S. climate refugees. But should the Biloxi-Chitimacha-Choctaw simply be designated as "climate migrants" or "climate refugees"? What about the history of colonialism that displaced the tribe to the coastal island during the 1830s, when genocidal settler militias and the Indian Removal Act displaced them and other Indigenous communities of the Gulf Coast?

These questions multiply in situations in which communities that have experienced displacement, segregation, and economic marginalization prior to the great acceleration have been concentrated in areas that are vulnerable to rapid environmental change. In such locations, structural factors ranging from zoning policies to insurance restrictions directly affect the potential for individuals and communities to stay in place and rebuild after disaster. The application of the word "climate" as an adjective that modifies "migrant" or "refugee" follows earlier attempts by the United States and other refugee-receiving states to parse differences between "political" refugees and "economic" migrants. During the Cold War, authorized in the programs of both major political parties, the U.S. government claimed to welcome refugees fleeing political persecution, but it dismissed migrants whom it viewed as fleeing economic hardship as ineligible for refugee status. U.S. immigration authorities relied on Cold War imperial mappings of capitalist versus communist states as the yardstick to measure which immigrants were deserving of refugee status and which were not. The U.S. Coast Guard carried out the refoulement, or forcible return, of migrants from

Haiti fleeing the rule of both Duvalier in the 1980s and Cédras in the early 1990s. Haitians were dismissed as economic migrants despite clear evidence of the political targeting of dissidents, while the United States allowed and even aided the transport of Cuban migrants automatically designated as political refugees. Such decisions elsewhere also mirrored the capitalist ally versus communist enemy logic; Mexicans were rejected as economic migrants while Vietnamese were at times aided in relocation, designated as political refugees. According to this Cold War logic, it was inconceivable that people entering the United States from capitalist countries would be categorized as refugees.[20]

Population Bombs and Environmental Exodus

An attempt to synthesize environmental links to political and economic factors is apparent in the first in-depth study of climate migration by an international environmental NGO. In 1995, Norman Myers and Jennifer Kent, under the auspices of the Washington-based Climate Institute, published the 214-page report *Environmental Exodus: An Emergent Crisis in the Global Arena*. Arguing that "environmental refugees"—people involuntarily displaced by climate change, weather disasters, drought, crop failure, soil degradation, and other processes—had grown so quickly since the 1980s that they numbered higher than traditional political refugees, the text outlines a policy agenda to both prevent mass environmental migration and address the outcomes of what they describe as the "emergent crisis." *Environmental Exodus* argues that climate changes and the collapse of agriculture risk displacing hundreds of millions of precarious, poor, and rural people in the Global South. Yet the text also vacillates between distinguishing environmental from economic migrants and intertwining economic and environmental causes of migration. Departing from el-Hannawi's insistence on the interrelation of infrastructure, development, and environment, this tension is resolved in *Environmental Exodus* by situating economic influences as contributing factors in migration decisions while giving environmental factors primacy as a "necessary cause" of migration. Claiming without evidence that economic attraction is often overstated as a migration pull factor and that otherwise, "many more Mexicans would have made their way to the United States," the authors assert that "neither environmental push nor economic pull need be a wholly sufficient cause of migration—though in the cases examined in this report we shall find that environmental factors are a necessary cause. It is not strictly relevant to consider whether in any specific instance they are prominent, pre-eminent, or predominant, though this report generally assumes that they are critical because they are fundamental."[21]

This assumption that environmental causes play a critical role that often overshadows economic factors—embedded in the methodology of the report—appears again in a discussion of the Haitian migrations of the Duvalier and Cédras years. The report asserts not only that political repression, poverty, and environmental destruction intertwine but also that the environmental factors exist independently of the political economy:

> There has been much repression from a series of authoritarian governments, and there is gross maldistribution of wealth and income. . . . It is difficult of course to separate out these political and economic factors from environmental problems, and hence to determine which is the more potent source of Haiti's deprivation. The two sets of determinants are closely intertwined, and they probably exert a compounding impact on one another. All the same, recall that the great bulk of citizens rely on agriculture for their living, and that the agricultural resource base has been severely depleted by environmental problems. . . . The situation is exacerbated by population growth of 2.3 percent per year, the second highest rate in the Caribbean. Even were a democratic government to be permanently restored, were there to be a more egalitarian society and economy, and were agrarian reform to be widely implemented—all of these being policy purposes of the US intervention—there would still be environmental problems aplenty that would keep large numbers of Haitians in absolute poverty for the foreseeable future. To a substantial extent, people fleeing Haiti can be considered primarily environmental refugees, though often with political repression at work too.[22]

Again anticipating some contemporary discourses on climate refugees, this passage indicates the ways in which political and economic determinants of migration may be retrospectively interpreted as the result of environmental processes, displacing attention to the ideological grievances of activists challenging the authoritarian state and thus subject to political repression. The text configures the hundreds of thousands of Haitian migrants to the United States as significantly driven by lack of adequate arable land in Haiti, despite the fact that the political repression, economic collapse, and repeated European and U.S. interventions and economic warfare created short- and long-term bases for outmigration in the late 1980s. In the process, it subtly marks environmental refugees as agents—not just victims—of environmental destruction. The authors assert the mutually reinforcing nature of economic deprivation and environmental degradation, suggesting that as impoverished agrarian populations expand, the "population pressure" they exert on natural systems drives migration to marginal lands, where the new arrivals tend to strip the soil of productive capacity and further destroy the land base. Elsewhere, the report makes these connections explicit, figuring environmental refugees as "destitutes" who degrade the environment: "It is the most impoverished who do the most environmental damage, however unintentionally."[23]

This foundational text unveiling the concept of the climate refugee to international policy audiences thus figures environmental refugees as both tragic victims and hapless agents in processes of environmental destruction. This builds on some of the long-standing contradictions within colonial and anti-Black depictions of peasants—especially rural peoples in Africa—connecting tropes of overpopulation with romantic invocations of environmental tragedy. In these ways, *Environmental Exodus* makes explicit its connection to the ideas of English cleric and demographer Thomas Malthus, whose deterministic arguments about the effects of overpopulation in the 1798 book *An Essay on the Principle of Population* suggest the propensity for the poor to overwhelm the so-called carrying capacity of land and resources in the society, a phenomenon euphemized in environmental discourse as "population pressure." For Myers and Kent, "Environmental refugees can sometimes be viewed, in part at least, as 'population pressure' refugees."[24] Malthus's ideas tend to portray the poor as the agents of their own misery due to excess reproduction. This is a line of argument that U.S.-based neo-Malthusian environmentalists in the 1970s and 1980s rearticulated in a global context in which the purported overpopulation of poor countries was viewed as the potential source of economic degradation, food crisis, and geopolitical emergency. Following publication of Paul and Anne Ehrlich's *The Population Bomb*, this idea was also popularized by Lester Brown and his Worldwatch Institute, who configured environmental problems largely as third world population pressures. India and China were often the focus of such reporting as their populations neared one billion each.[25] Building on this focus of international environmentalists, Myers and Kent's ostensible humanitarian focus is coupled with a racially potent imaginary of southern population-driven environmental degradation. The ambivalence between these two poles of emphasis in the report is evident in the one photograph it contains, where the gendering of environmental

Norman Myers and Jennifer Kent, *Environmental Exodus: An Emergent Crisis in the Global Arena* (Washington, DC: Climate Institute, 1995).

degradation is explicit. Without any caption explaining the image, the cover of the report includes the portrait of a veiled Black woman holding a baby as they both stare directly into the camera. Echoing similar colonial portraiture of the colonized mother as an icon of the tragic social and environmental conditions of native societies in Africa and Asia—Katherine Mayo's image of an Indian mother and child in the 1927 book *Mother India* being one widely circulated example—the cover of the report deploys dark-skinned mothers and children as the conventional objects of colonial humanitarian rescue, even as the invocation of the child suggests the incidental potential for the marginalized environmental refugee to reproduce the same conditions of displacement through the mother's excess capacity for reproduction.[26]

Today, the legal and political context of the term "climate refugee" is different from the moment in which Myers and Kent published their report, especially as it emerges in two contexts that make migration an increasingly

Katherine Mayo, *Mother India* (New York: Blue Ribbon, 1927).

politicized crisis issue: (1) neoliberal economic policies, broadly dispersed armed conflict, and environmental disasters have generated massified flows of both internal and transborder migration, often fueled by demand for labor in the rich countries; and (2) the rise of neoconservative border militarization, coinciding with Islamophobic discourses on terrorism, has both limited the scope of formal migration pathways and grown the flows of officially unauthorized (though economically demanded) migration. As Nicholas De Genova argues, in the current moment "the criteria for granting asylum tend to be so stringent, so completely predicated on suspicion, that it is perfectly reasonable to contend that what asylum regimes really produce is a mass of purportedly 'bogus' asylum seekers." For De Genova, this scene or spectacle of illegality is always accompanied by its "obscene supplement: the large-scale recruitment of illegalized migrants as legally vulnerable, precarious, and thus tractable labor."[27] For this reason, Harsha Walia suggests that we reframe thinking about "border control" to recognize the realities of "border imperialism": "Border controls are most severely deployed *by* those Western regimes that create mass displacement, and are most severely deployed *against* those whose very recourse to migration results from the ravages of capital and military occupations."[28] As such, the widespread public media discourse that configured transnational migration to Europe as a "refugee crisis" in 2015 made the state itself the entity that experienced crisis rather than the people who were displaced.[29]

Given the imbalance between demand for transnational labor migration and the declining willingness of receiving states to authorize continued migration flows, it should be no surprise that the articulation of climate migration as a new juridical category of international migration law has often been led by climate-affected states rather than receiving states. In cases when receiving states (particularly in the European Union) advocate for greater quotas of environmental migrants, the case is usually made by reference to the potential security crisis of mass unauthorized migration. These diplomatic dynamics were present in negotiations on the 2018 Global Compact for Migration, when a number of climate-affected states and international NGOs advocated for language recognizing climate change's impact on migration. Despite the strategic goal of such advocacy in allowing larger numbers of legal entries, the calls for developing regular categories in international law for climate-affected migrants create a context in which states attempt to designate some migrants as deserving of relief over others because of the environmental roots of their mobility. One strategic reason climate-affected states might advocate for "climate refugee" as a new category within international law is precisely because it bypasses broader questions about the continuing inequalities in the international system, which are exacerbated by carbon extraction and the climate change processes it fuels.

If the originating accounts of environmental and climate refugees that came out of environmental NGO policy studies reproduce conservative narratives of degradation that blame victims of disaster for their own plight, other more recent environmental discourses about human-induced climate change tend to move toward apocalyptic predictions about the scale of such degradation. In the move from the degradation narratives of the 1990s to the apocalyptic visions of climate change in the twenty-first century, we glimpse how climate migration discourses enable a growing focus on the waste products of the system of racial capitalism.

Colonial Mirror

Climate Migration as a Legal Concern

If in the 1990s climate migration discourse began to articulate an environmental source of global insecurity, by the 2000s and 2010s a variety of human rights, legal, and environmental justice discourses began to advocate for humanitarian responses to climate migration. This became clear in the legal case of Ioane Teitiota, a man from Kiribati who sought to be legally declared the world's first climate refugee in a petition to the government of New Zealand. Teitiota migrated with his wife, Angua Erika, in 2007 from South Tarawa, Kiribati, to New Zealand, where he worked as a farm laborer. The couple had three children. After overstaying his visa, Teitiota applied for asylum under the 1951 Geneva Convention Relating to the Status of Refugees, which is incorporated into New Zealand's 2009 Immigration Act. When the petition prepared by his pro bono lawyer, the Pentecostal minister Michael Kidd, was rejected by New Zealand's Immigration and Refugee Tribunal on July 20, 2015, the court declared that its decision should not be taken as general precedent; the decision does "not mean that environmental degradation resulting from climate change or other natural disasters could never create a pathway into the Refugee Convention or protected person jurisdiction."[30] Although Teitiota's deportation on September 23, 2015, and failure to win a final appeal hearing in the Supreme Court meant that he was permanently barred from New Zealand, he subsequently took the case to the United Nations. Again, the UN Human Rights Committee upheld New Zealand's deportation decision. However, a variety of international media outlets immediately reported on the ruling, which determined, according to Amnesty International, "that governments must take into account the human rights violations caused by the climate crisis when considering deportation of asylum seekers."[31] *Time* magazine's headline indicated further that "climate refugees cannot be forced home."[32]

The involvement of the UN Human Rights Committee and the statements by Amnesty International reflect the increasing participation of human rights organizations in climate change as a legal and political matter. These organizations tend to foreground migration as a significant signal of human rights violations stemming from climate insecurity. In 2009, Amnesty International began calling for emissions reductions as an international human rights concern and indicated that if polluter states do not take action, they will "bear responsibility for loss of life and other human rights violations and abuses on an unprecedented scale."[33] Other major human rights NGOs including Human Rights Watch have moved into climate advocacy over the past decade. The 2019 report *Climate Change and Poverty*, released by Philip Alston, an international law scholar and United Nations special rapporteur on extreme poverty and human rights, criticizes the complacency of human rights bodies and NGOs, as well as the complicity of corporations and governments around the world, with fossil-fueled climate change. Noting that "forced migration," or the choice "between starvation and migration," is one of the likely large-scale outcomes of climate change, Alston argues that structural transformation of the world economy, litigation, and broad-based community solutions to climate change impacts are necessary to offset massive job losses and housing and food crises affecting the world's poor.[34]

From Alston's vantage within the human rights community, the aforementioned public reports about the UN tribunal ruling might appear irresponsibly celebratory. *Time*'s headline declaring that climate refugees cannot be returned does not address the complexity of the New Zealand tribunal and UN committee decisions, which both suggest at minimum an uncertain threshold for when settler states will accept asylum claims—or worse, that the threshold for being protected as a climate refugee is unlikely to apply to the displaced until their home countries are literally unable to support human life. Both decisions draw on a number of expert discourses on climate change. Because the convention guarantees a "right to life," both legal bodies addressed expert testimony and research documents that examined the changing ecological situation generally among the thirty-three islands of Kiribati and specifically concerning Teitiota's place of residence, Tarawa, which is home to nearly half of the I-Kiribati population. In Tarawa, displacements from other islands are concentrating the population; as Tarawa's own land base and water table shrinks due to pollution and sea level rise, land disputes have become increasingly common and contentious. Researchers reported malnutrition, disease, crop failure, and water shortages, compounded by mass unemployment and widespread reliance on subsistence agriculture. These are indicators of a broader assessment attributed by the UN committee to the Intergovernmental Panel on Climate Change from its fifth assessment report, released in November 2014: "Kiribati . . . is losing land mass and can be expected to survive as a country for 10 to 15 more years."[35] And yet this period of national habitability is apparently enough in the eyes of the UN committee to ensure that Teitiota's forced return to Kiribati remains legal at present; because Teitiota would not be immediately threatened with death, he could be returned by New Zealand authorities to a country configured by expert discourse on climate change as inhabiting a shrinking future.

Tim McDonald, "The Man Who Would Be the First Climate Refugee," BBC News, November 5, 2015, www.bbc.com/news/world-asia-34674374.

Kidd's publicity for the case helped ensure that the New Zealand tribunal decision would be watched internationally, with televised coverage from Australia to France, as the site of a "test case" on the climate refugee; knowing that the law itself is unlikely to provide remedy, Kidd proclaims, "The law needs to be changed."[36] In this context, potential refugees like Teitiota have the ability to tell their stories to broader audiences, but the narrative frame of these stories is constrained in important ways. The context of refugee law requires that refugees strategically lay claim to stories that reify the receiving state as the last line of defense against impending death. These are thus narratives configured by national dependency; in the words of Epeli Hau'ofa, "Is this not what neocolonialism is all about? To make people believe they have no choice but to depend?"[37] Teitiota's own interviews with Western journalists confirm this vision of Kiribati as a kind of "deathworld," configured in order to underline a (now failed) humanitarian claim to refugee status.[38] In an online video interview with the BBC, Teitiota is shown alongside his children at home in Tarawa after his deportation, indicating places where the flood wall has been damaged and the towering height that the tide reaches at its zenith. After displaying welts on his children's skin to show that they have been affected by malnutrition and poor water access since their return from New Zealand, Teitiota indicates through a translator that his situation is "the same as people who are fleeing war. Those who are afraid of dying, it's the same as me." Bringing together a vision of climate-generated vulnerability uniting present disability and future death, Teitiota's testimony underlines the failure of the potential receiving state of New Zealand to fulfill the promise of asylum and the certainty of extinction. Comparing himself and his family to refugees fleeing the direct violence of war, Teitiota states, "The sea level is coming up, and I will die, like them."[39]

The Specter of National Extinction

The consequences that flow from the speculative declaration of the demise of an entire national population cannot be delimited to the arena of law, of strategic claims migrants might deploy to gain entry into a receiving state. Confirming long-standing stereotypes of the inevitability of Indigenous disappearance,[40] the depiction of islands shrinking into rising seas looks somewhat different from public discourses concerning the threat of mass migrations from Syria, North Africa, or South Asia, where climate change mobilizes fears of swarming Islamist militancy amidst resource scarcity. As Renisa Mawani argues, the ocean has been configured by colonial powers as a free space for exploration, trade, and conquest even as it has simultaneously been rendered a site of repeated attempts to control population movement and enforced racialized laws separating colonial peripheries from European metropoles and white settler colonies.[41] The colonial imaginary of the sea as a site of individual freedom buttresses visions of the island as prison, a historical situation that reinforces the reported I-Kiribati disdain for Teitiota's representation of the national population as refugees who must flee a vanishing island. As Suvendrini Perera thus argues, renewed interest in the seafaring traditions in Indigenous Pacific literature and public culture often grapples with the complexity of violence, economic hardship, and environmental change that surround island refugee subjects.[42]

In the neocolonial vision of shrinking land under climate change, the predicted climate disaster that depopulates Kiribati recasts social time as shrinking at the individual, household, and national levels. The usual forms of time management that lend an imagined predictability to living—the life span of the individual, family time of reproduction, labor time based on industrial or crop cycles, and the indefinite opening of state time—all come under pressure with the stereotyped vision of climate change as a localized deathworld.[43] Built on a history of colonial representations of islands that, as Elizabeth DeLoughrey argues, are presented as metonymic for a global condition, the shrinking island as a deathworld of climate change allows for the depiction of environmental determinism precisely because the assumption of a premodern past presages a deathly future.[44]

The change in the type of imaginable time for living inevitably bleeds into the experience of the everyday—a "descent into the ordinary"[45]—affecting the interpretation of life events along the slow-moving lines of environmental disaster. In snapshot moments, this bleeding of crisis feeling into the interpretation of social and ecological change may appear to transition a person, a family, a community, or an entire nation across the threshold separating the creeping motion of "slow violence" upon a lifeworld to the rapid undoing of life itself in "crisis."[46] What does the specter of mass displacement do to transform the feeling of inhabiting the present? How does it enable certain visions of what constitutes a livable life or the shrinking horizon of the future? To the extent that the specter of the climate migrant haunts transnational environmental discourse and immigration law, it configures the migrant's lifeworld, generically conceived as an individual or a familial struggle rather than one at the scale of community, nation, or planet, as a compressed and constricted ecology requiring intervention. It is this potential for securitization—woven into the temporal unfolding of crisis from slow to rapid onset—that both views migrant bodies as inhabiting a space of living death and animates a kind of government of the climate migrant's life along lines anticipated by discourses of asylum, adaptation, and development. Migration thus becomes not only an object to be contained by security but also a resource of adaptation to climate change.

In such a movement from a deathworld to a displaced migrant futurism, Aimee Bahng recognizes the speculative capture of migrant bodies as projections "that materialize the abstract, rendering it available for possession."[47] Herein is a specific and newly emergent process of race-making endemic to climate migration law as it attempts to move from individual claimants to narratives of group vulnerability using the markers of corporeal, biological, or geographic difference. This is evident in the manner by which courts distinguish differential risk to specific groups in the adjudication of the handful of extant climate-related asylum claims. Despite the fact that Teitiota's case was widely reported as the first legal attempt to establish climate refugee status in law, the New Zealand decision against Teitiota cites a history of other similar petitioners—all originating from Tuvalu—dating to the year 2000. In the first of these cases, rejected upon appeal on August 10, 2000, an unnamed seventeen-year-old Tuvaluan man was denied asylum because all Tuvaluans were subject to displacement, and he was not considered by the court to be differentially affected as a member of a protected group within Tuvalu. The fact that refugee law generally and the Geneva Convention specifically are dependent on a narrative of minority persecution within a nation-state allows the court to dismiss Tuvaluan claims for asylum precisely because of their universal vulnerability to climate change:

> None of the fears articulated by the appellant *vis-à-vis* his return to Tuvalu, can be said to be *for reason of* any one of the five Convention grounds in terms of the Refugee Convention, namely race, religion, nationality, membership of a particular group and political opinion. This is not a case where the appellant can be said to be differentially at risk of harm amounting to persecution due to any one of these five grounds. All Tuvalu citizens face the same environmental problems and economic difficulties living in Tuvalu. Rather, the appellant is an unfortunate victim, like all other Tuvaluan citizens, of the forces of nature leading to the erosion of coastland and the family home being partially submerged at high tide. As for the shortage of drinkable water and medicines, medical care, doctors and other associated services, these are also deficiencies in the social services of Tuvalu that apply indiscriminately to all citizens of Tuvalu and cannot be said to be forms of harm directed at the appellant for reason of his civil or political status.[48]

need to reconceptualize what "minority" is

It is precisely in explaining the court's denial that climate change conventionally discriminates based on race, nationality, or other protected group that a judgment is made that meets Ruth Wilson Gilmore's definition of structural racism as "state-sanctioned or extralegal group differentiated vulnerability to premature death."[49] Here the court collects people into a group differentiated by their shared vulnerability to premature death. Race-making thus takes the form of collecting an entire national population subjected to escalating ecological violence and configuring their shrinking horizon of livability as an incidental process rather than targeted discrimination. Even though the implication is that a vulnerable national population would need to eventually be reconfigured as a vulnerable minority in diaspora, such a designation disallows the law from considering Tuvaluans as a targeted group. Put simply by the court, "all" Tuvaluans are "unfortunate victim[s]." This casual representation of the extinction of an entire national population as an unintentional tragedy cynically masks the combination of environmental, economic, and social pressures that render small islands especially vulnerable to climate change. Despite the fact that the category

"Tuvaluan" is repeatedly used to identify the vulnerable population of which the petitioner is a part, the environmental racism of the wealthy polluter nations is not grounds to understand this as a threat of death emanating from protected-category discrimination. Instead, the abstract agent of climate destruction is "forces of nature." As such, the petitioner cannot seek remedy of asylum from New Zealand, and because of Tuvalu's very representation as an undifferentiated deathworld in law, the court condemns the petitioner to return to the place it views as having no future. In the logic of New Zealand's adjudication of the Geneva Convention, if minoritization by ethnic cleansing supposedly entitles targeted migrants to the mythic "gift" of passage advertised by receiving states,[50] universal vulnerability to the slow violence of climate change ensures their abandonment to enclosure, to a carceral logic of exclusion upon shrinking islands of life.

Migration, Human Security, and Environmental Racism

Although climate migration discourse has thus far resulted in little redress or relief for migrants, it has helped to advertise environmental inequalities. In turn, climate migration discourses have begun to appropriate the language of environmental justice movements, even as such language may sit uneasily beside the global crisis discourses and security logics that suffuse public reporting about migration. Environmental justice studies has thus occupied a complex position both inside and adjacent to climate migration discourses. Pioneered by grassroots environmental activists, environmental justice research in the tradition of sociologist Robert Bullard analyzed the siting of waste near Black communities in the United States beginning in the 1980s. This sociological research in the U.S. context worked to understand how the risks of toxicity emerged out of historical forms of segregation and were geographically concentrated in ways that unequally affect Black and poor communities. Environmental justice scholarship building on this tradition, including key writings by Laura Pulido, Julie Sze, David Pellow, and others, has since taken up a variety of themes, ranging from toxic air pollution, waste disposal and recycling, and the effects of industrial processes and pesticides on workers to environmental justice protest and questions of urban space and design.[51] Such work offers important pathways for thinking about widely distributed environmental changes because it grapples with questions of responsibility and inequality. If corporate polluters mask the exact harms they unleash when they pollute environments, how can their culpability be proven? And how do invocations of environmental inequalities relate to recent efforts to incorporate migration and security discourses into environmental knowledge?

The appropriation of environmental justice concepts is evident in discussions among climate researchers and the IPCC of how migration itself may be one of the key acts of resilience by climate-affected peoples. In such claims, public assessments of climate change as racism often invoke retrospective empirical calculation of racialized groups' vulnerability to environmental destruction—a configuration of racial power that may be incidental in intent but deadly in effect. In this context in which dominant environmental governance considers racism an incidental effect of climate disaster, academic research and policy proposals concerning climate change increasingly invoke paradigms of security that claim to address environmental racism and displacement.

Environmental racism is in turn posited as a problem for thinking about human security, as it may both signal the vulnerability that human groups experience due to environmental change and the capacities they exhibit for resilience. The term "human security" operates as a baseline to consider how climate-affected peoples interface with a broad range of social, political, economic, and environmental systems. Purportedly a counter to militarized security emerging from development studies in the 1990s, the concept represents a liberal approach to security that in the post-9/11 era integrates environmental, biological, and social systems into the surveillance, data gathering, and intervention processes of state and interstate institutions. Especially popular in the European Union context as well as among international NGOs, "human security" suggests that humans share universal needs and are subjected to various types of risks that can be the objects of prediction and intervention. Despite the apparent internationalism of human security, which claims to transcend narrow national security agendas, this concept has a wide arena of application and has recently been deployed as a form of governance integrating security and humanitarianism. Combining network analysis, surveillance, policing, military intervention, and the statistical management of populations, economies, and environments, human security activates a "posthuman" form of governance that is premised on creating ever more detailed data forms for predicting potential insecurity and devising counterstrategies. Although this characterization differs from some of the celebratory announcements of human security in the 1990s as either "freedom from fear" or "freedom from want," the integration of surveillance, intervention, and network analysis renders some contemporary contexts of human security practice as extensions of existing foreign policy and national security agendas in a manner that reifies northern intervention as the site of the production of freedoms.[52] This corresponds to shifts away from questions of territorial sovereignty in security discourse and toward the forms of informational control that have recently become significant to understanding digital governance and warfare.[53] In this modality of control, bodies are targeted not primarily through techniques of inclusion and exclusion or through subjectivation but rather through the calculation of gradations of difference in population constructions.[54] Invoking a network model relating social groups to technical, environmental, and geopolitical systems, human security engages in predictive forms of modeling that require aggregated social categories (including racial categories) in order to construct risk differentials.[55] In this type of security thinking, environmental racism—represented in terms of unequal vulnerability to disaster—represents an increasingly useful tool for surveillance and modeling of large-scale social problems.

In the fifth assessment report of the Intergovernmental Panel on Climate Change, the chapter on "human security" notes differential economic and racial factors that influence the possibilities for the return or resettlement of environmental migrants. Even as risk is intensified by weather-driven crises, resources for resettlement create wide disparities in the IPCC's rendering of human security:

> Most displaced people attempt to return to their original residence and rebuild as soon as is practicable. The Pakistan floods of 2010, for example, caused primarily localized displacement for large numbers of people across a wide area, rather than longer-distance migration. Structural economic causes of social vulnerability may determine whether temporary displacement turns into permanent migration. In New Orleans, after Hurricane Katrina, for example, economically disadvantaged populations were displaced in the immediate aftermath and most have not returned. . . . Fourteen months after the event, African American residents returned more slowly, because they had suffered greater housing damage. Studies conclude that displacement affected human security through housing, economic, and health outcomes and that these have perpetuated the initial impact into a chronic syndrome of insecurity.[56]

The IPCC's formulation of climate change as a threat to human security links the effects of displacement to racialized potentials for precaritization and rehabilitation. On the one hand, the figure of the climate refugee is marked by a racialized chain of waste effects, as carbon emissions render lands transitory and intermittently productive. At the same time, the climate refugee's exercise of mobility is itself a resource against the potential catastrophe of spectacular climate disasters that happen at the shorelines of the desert and sea. Detached from the scene of disaster and returned to settled land elsewhere, the recapacitated climate refugee is rendered as a site of hope for the resilience of populations in the aftermath of disaster. The IPCC refers to historical scholarship on "the relationship between large-scale disruptions in climate and the collapse of past empires" as evidence of climate's security threat (rather than emphasizing the reverse proposition that empire threatens life itself through climate change).[57] Thus, even as such discourse positions the colonial state as the arbiter of security and conservation, the figure of the climate refugee ideologically functions as a supplement to the colonial geopolitical order, a positioning that allows for the valorization of the refugee's purported knowledges about and practices of "climate adaptation." Forms of knowledge produced in displacement are increasingly fetishized as Indigenous resources for climate resilience, although these processes vary regionally and are contested, for example, in the emergent Latin American "rights of nature" debates.[58] In rendering racialized environmental capacities as "human security," adaptive forms of flexible migratory living portend the future of capital after climate disaster. Such ideas often confer a kind of romanticism to the climate migrant, who is first a security risk but who may be rehabilitated in the service of universal human adaptation. In the process, references to human security in climate research and policy appropriate environmental justice discourse to suggest widely dispersed inequalities of disaster while suggesting that the Indigenous knowledges of migrants and other climate-affected peoples are resources that can be scaled up and redeployed across different geographies and scales of disaster.[59] Race is thus a vital aspect of representation of the climate migrant, whose difference is mobilized in order to represent both the current harms of and the potential for future adaptations to climate change.

If the incorporation of environmental justice concepts into climate migration discourse has been a fraught exercise, revealing that the nexus of humanitarian and security thinking might lead to new forms of racialized population construction around weather-displaced migrants, how might environmental justice critiques be redirected to critically analyze the security discourses with which they have been entangled? The manner in which differences of race, class, and nation are calculated as vulnerabilities to unequal environmental harm is thus being used to invigorate new forms of climate migration governance. A turn toward the critical examination of key concepts within the field of environmental justice studies is necessary on this point, as scholars begin to expand the scope of what is considered environmental injustice and think critically about how social and geopolitical factors interrelate with phenomena seen as "environmental." In the context of the dual pandemics of COVID-19 and police violence that activists are challenging at the present moment, such linkages remain critical.[60] Although environmental justice scholarship has made vital contributions to understanding "the unequal distribution of environmental benefits and pollution burdens based on race,"[61] the field's historic and important focus on ex post facto determinations of unequal vulnerability to environmental harm may have had the unintended effect of narrowing the field of analysis of environmental racism on three fronts: it excludes the productive bases of racial capitalism, it repeatedly stages a debate over whether race or class is a greater contributor to toxicity risk, and it sidelines attention to processes of colonial settlement that displace alternatives to capitalist forms of development.[62] So although it remains necessary to maintain focus on racially unequal distributions of environmental harm, invocations of those inequalities must be a starting point rather than an end point of the analysis. It has been critical to environmental justice research that pollution siting is not incidental, that injustices are structurally reproduced. Ex post facto evidence of environmental racism—such as discussions of how toxic waste siting unequally affects communities of color in the United States—thus benefits from being connected to an analysis of broader economic and political currents that invest in and reproduce inequality, including the logics of security that affect climate thinking today.

The turns to the incidental and the empirical in discussions of environmental racism invoke an ad hoc conception of racism. What would it mean to think, along the lines of Arun Saldanha, of climate change more broadly as a racial ecology of capitalism? Dismissing stereotypes of poor and Indigenous peoples as being the agents of degradation, for Saldanha, "the ecology of global capitalism has for some four centuries been intrinsically *racist*, making white populations live longer and better at the expense of the toil and suffering of others. Humanitarian campaigns after 'natural' disasters in the South (the 2010 Haiti earthquake), disasters which will become routine if capitalism goes on as it does, are the clearest example of the continuing racist hypocrisy underneath Western humanism. . . . As activists point out, places suffering most from climate change have contributed least to carbon emissions. The Anthropocene is in itself a racist biopolitical reality."[63] To challenge the empirical inequalities outlined in environmental justice research, then, requires attention to their mobilization in systems of racial control, to the ways that organizations like the IPCC can mobilize data points on climate injustice toward new models of governance.

Such an approach would follow important research on the relationship between race and capitalism. Ruth Wilson Gilmore's study *Golden Gulag* explores how neoliberal prison expansion in California serves as a solution to crises of political economy, assembling racial categories by reproducing and redistributing social vulnerability. It is in this context—in which the criminalization of populations figured as surplus resolves contradictions in the state's management of capital's development potential—that Gilmore defines racism as an institutional social relation: "Racism, specifically, is the state-sanctioned or extralegal production and exploitation of group-differentiated vulnerability to premature death."[64] From the vantage of the neoconservative common sense about race, as well as from class-centered forms of leftist critique, this definition may seem idiosyncratic or simply inaccurate. Centering on the distribution of death rather than the differentiation of life (via phenotype or cultural essence), this definition suggests that race is the product of social relations rather than an a priori social categorization. The potential of such a redefinition of race lies in part in the ways it registers what Nikhil Singh describes as "the apparent contradiction between the ongoing normalization of racial liberalism and the intensification of racially inscribed domination."[65] Racism must be understood not simply in its rhetorical form as a set of moral infractions but rather as an effect of the material formation of social relations and their imbrication in more-than-human networks of settlement and ecological reproduction.

Thus, despite the fact that Gilmore's focus on racism's relation to death shares a certain conceptual ground with dehistoricized theories of race as social death or exception, her insistence that the social dynamics of this death-dealing relation are *productive* of race rather than its *effect* represents a significant break with these other methods. This is evident in Gilmore's brief discussion of Islamophobia as at once a reproduction of racial hierarchy and a transition:

> Sadly, even activists committed to antiracist organizing renovate commonsense divisions by objectifying certain kinds of people into a pre-given category that then automatically gets oppressed. What's the alternative? To see how the very capacities we struggle to turn to other purposes *make* races by making some people, and their biological and

> fictive kin, vulnerable to forces that make premature death likely and in some ways distinctive. The racialization of Muslims in the current era does double duty in both establishing an enemy whose being can be projected through the allegation of unshakeable heritage (fundamentally, what the fiction of race is at best) and renewing the racial order of the US polity as normal, even as it changes.[66]

As one of the energizing forces driving security thinking, the post-9/11 racialization of Muslims as civilizational threats in the northern countries is one condition for the promulgation of a climate security discourse that configures migration as a threat to the state system—particularly given the number of Muslim-majority states that are depicted as climate migration hot spots. This dynamic will be discussed further in chapters 3 and 4. Racism, as it integrates migrants into security frameworks, is not the effect of race, dependent on its prior differentiation. Racism instead structures social forces that reconfigure populations and their racialized contours through embodied interactions and collective struggles. Drawing from Gilmore's assessment of the relation between structural vulnerability and the reproduction of racism, critical scholarship on environmental racism can benefit from analyzing how the domain of empirical environmental inequalities feeds back into the productive forces of capitalism, a concern I explore at length in chapter 2. As the contemporary politics of both anti-Muslim racism and environmental racism demonstrate, retrospective figurations of racial disparity are persistently disavowed by actors who reframe race via discourses of security or development.

Conclusions

In sum, this chapter has argued that the public figure of the climate migrant helps to advertise both the widely distributed risks and the potentials for securitization that emerge in the post-Kyoto era of rapid climate change and state-led environmental adaptation. By advertising a racialized and gendered vulnerability of the climate migrant to ecological forces, climate security discourse imagines rural peoples in climate-affected areas as tragic victims of first world environmental processes, whose debilities require external intervention. At the same time, the climate migrant is configured as likely to accelerate the environmental degradation, risking a cascade of migration and potential violence that threatens to undo the normative powers of the nation-state to control movement and maintain infrastructure. As such, the climate migrant presages the racialized potential for climate change to undo the power of sovereignty, which recapitulates Malthusian narratives of resource conflict, energizes legal processes that continue to confirm the power of the state to exclude migrants, and activates liberal forms of racial control that appropriate environmental justice concepts to serve the banner of human security.

In contrast to this chain of risk speculations and security interventions, chapter 2 offers a theory of racial capitalism that explains the centrality of oil to the joint development of global crises (in war, environment, and economy) and the resulting labor transnationalization. Arguing that carbon-fueled development both radically increased global carbon emissions and set in motion new structures of capitalist resource use, the chapter contends that climatic factors in migration build on these existing processes of capitalist expansion, which have fueled transnational migration between Asian nations. I discuss in greater detail how the systems of finance that emerged after the 1970s oil boom created a racialized labor system that both divided southern states along intensifying economic inequalities and sustained new projects of United States empire through the end of the twentieth century. Highlighting how contemporary wars and the structure of U.S. economic power are dependent on oil, the chapter explains how racialized migration panics around terrorism and climate change have emerged from the intersection of oil and finance in the international economy, including in recent turns to "green finance." As such, xenophobic fears of climate migration to the North are a symptom of a broader restructuring of the neoliberal economy that makes both migration and remittances into strategic development and accumulation strategies in the Global South. Although economic and political factors are central in the massification of migration flows under neoliberalism, environmental processes may contribute to such factors, requiring a theory of race that links structures of extraction, labor, and finance to broader ecologies of social reproduction.

CHAPTER TWO

The Changing Wealth of Nations

Oil, Labor, and Racial Capitalism

Although the figure of the climate migrant as a potential security problem increasingly appears across continents—in public depictions of weather disasters from the Horn of Africa to Central America, from island nations in the Indian Ocean to the U.S. Gulf Coast—some of the key migration pathways established for the era of the so-called great acceleration in fossil fuel emissions emerged after the 1973 Gulf oil crisis transformed international relations and finance. This chapter tells the story of oil's restructuring of geopolitics with an eye to the ways that the resulting shifts in labor migration routes across Asia played a role in developing racialized structures of labor in advance of intensified and rapid climate changes. In the process, the chapter argues for a materialist understanding of the links between environmental and economic bases of migration, and offers a method for tracing the interrelation of oil, capitalism, state policies, and human migration as an alternative to the often deterministic or even apocalyptic visions of environmental change in climate migration discourse. As such, in order to understand the relation of climate change and migration, I ask the reader to follow key developments in international relations and political economy, which are fundamental to understanding the racial character of oil's role in twenty-first-century migration patterns. Based on this chapter's analysis of racial capitalism and the migration flows it produces, subsequent chapters will return to a critique of racialized discourses about climate migration in Bangladesh and climate war in Syria.

The Gulf oil shock is one of the critical historical events that will help us center the history of oil in the development of contemporary racial capitalism's violent economic and environmental effects. This event involved a sudden spike in the price of oil on the international market and is one of the better-known episodes in the world history of the energy economy. When oil-producing countries placed a brief embargo on oil exports in 1973 in protest of northern support for Israel in the October War, the resulting petroleum shortages in the United States and the price spike on the world market transformed the structure of international finance and reoriented political relationships among oil-producing states, oil-dependent southern nations, and oil-consuming economic powers such as Japan and the United States. However, it is not generally mentioned by scholars of economic history or political economy that this oil shock—conventionally framed as an economic shift, a geopolitical crisis, and, more recently, a sign of destructive carbon consumption—had a specifically racial character, and that its influence on development strategies and transnational labor relationships was significant to the development of contemporary crises over migration. Building on theories of racial capitalism that track how race is structurally reproduced in the rescaling and expansion of different forms of capitalist networking, I argue in this chapter that race plays important roles in the history of Gulf oil, not only through direct racism in the labor structure of the oil fields but also in the manner in which oil-based finance and development has generated new forms of migration in Asian nations that have become central to international systems of capitalist production and trade. Such migration commonly works along two pathways: (1) from agriculture-dependent rural or coastal areas to cities, and (2) from poorer countries in South and Southeast Asia to wealthier ones in West Asia. The development of the South Asia–Persian Gulf migration corridor, discussed further in chapter 3, is one of the main historical outgrowths of this transformation—a development that has expanded a racially stratified and gender-segmented labor market in the Gulf states, following the logics of earlier colonial mass labor accumulation strategies, such as the British and Dutch "coolie" trades. In the process, the geographic bases for the development of Asian diasporas in the Gulf states and for some of the Asian populations moving northward in the so-called European migrant crisis were generated through oil's restructuring of inter-Asian capitalism. Weather disasters today often move migrants along the pathways established by oil in such migration corridors, even as changing geographies of oil extraction and new forms of climate adaptation and green capitalism create the potential for shifting or multiplying these routes.

In sum, this chapter argues that by employing frameworks from critical migration studies and racial capitalism research to revisit the history of Gulf oil development, it is possible to understand the story of the rise of neoliberalism as planetary capitalist logic and its evolving configuration of racialized migration politics. In the process, my analysis focuses on how the oil shock put the world economy on a path toward the three interrelated crises of deepening international debt, growing transborder migration, and intensifying anthropogenic climate change by the early twenty-first century. These transformations were felt most deeply among agrarian populations in debt-dependent poor countries, who were pushed into neoliberal development strategies that today contribute to significant displacement conceived as "climate migration."

By tracking the relation of race, oil, and environmental change, we can also come to understand some key transformations among the geopolitical formations connecting West and South Asia to northern states that have historically dominated international political and financial systems. We can begin to see how the neoliberal phase of racial capitalism—as a system reliant on the simultaneous externalization of carbon emissions and internalization of oil as an asset valued on its financial as well as physical capacities—has enabled a seemingly contradictory transformation of U.S. power in the world system. While U.S. wealth and power has been decentered internationally by U.S. dependencies on Gulf oil and Asian manufacturing—each of which produce massive new migrations within and across borders—the linkage of oil to the dollar in international trade has enabled the proliferation of debt-financed military expansion and international arms trade, which continue to reproduce U.S. military hegemony and state securitization, thus advancing the so-called war on terror alongside new initiatives in environmental security. As such, this chapter builds on existing scholarship that attempts to capture how "US wars in Asia, US racial capitalism, and US empire" have been deeply entangled in the making of contemporary geopolitics despite the fact that they remain "underrecognized parts of the genealogy of the contemporary condition."[1] Further, the chapter offers an alternative to mainstream journalism, policy, and security discourse on the figure of the climate refugee, arguing that the imperial relations of oil generate forms of racial displacement and vulnerability to violence, reflecting the productive role of surplus Asian labor in the maintenance of capitalism's systemic inequalities.

This chapter thus contributes to scholarship on race and migration by offering an account of the racial character of the energy system that has fueled neoliberalism, in the process connecting inter-Asian economic and labor shifts to longer histories of race and to the fates of marginalized peoples across the South. As research on racial capitalism originated in Black studies scholarship that worked to interpret the history of capitalism by attending to the Atlantic plantation complex, this chapter works to bridge earlier studies focusing on capital's accumulation of Black labor, with attention to the geographic shifts that make Asian countries central to labor, trade, and financial systems in the current moment. This requires a comparative framework that brings together the perspectives of Black studies, Asian studies, and Asian diaspora studies in order to understand both how the foundational anti-Black racism of capitalist expansion provides models for more recent systems of migrant labor exploitation and how today's inter-Asian geographies of capitalism proliferate forms of racialized displacement and disposability for many Black, Brown, and Indigenous peoples affected by climate change and international debt. This second aspect of racial capitalism is especially important, because even as large numbers of displaced agrarian people are recruited into neoliberal labor flows, many minorities and vulnerable workers have been relegated "superfluous," reflecting the attempts by wealthy countries to dislodge Black and other marginalized workers from the "labor-'light' economy of twenty-first century racial capitalism."[2]

Theorizing Racial Capitalism

Before we can understand how debt, oil, and inter-Asian migration are integrated into the racial logics of capitalism in the twenty-first century, it is necessary to understand how the intersection of race and capitalism has thus far been understood in the long time span of colonialism's development of transcontinental power dynamics. Studies of this relationship are not new, as they have long preoccupied scholars in a number of fields. Black studies research on the relationship of race and capitalism has a long lineage in the twentieth century, dating from early studies on the relationship of capitalism to slavery, war, and segregation to more recent works on prisons, policing, and urban capitalism.[3] A growing body of scholarship on the relationships of race, capitalism, and colonialism in critical race and ethnic studies and in North American Indigenous studies has developed new and important accounts of how expropriation of land and the spread of enslaved labor have been central to capitalism's systematic reproduction and geographic expansion.[4] Finally, discussions of the place of colonized nations in Africa, Asia, and the Americas within the history of the rise and development of capitalism—led by dependency theorists who describe neocolonial economic formations—have produced important debates about how colonial divisions in the international system affect sociological theories about capital.[5] Even if these latter studies focusing on the Global South do not explicitly invoke race as a key term for study, they deal with how categories of analysis of class and capitalism are implicitly racialized and conditioned by an international division of labor that works to exploit formerly colonized nations through the structures of trade, extraction, and development.

These three literatures engage with some of the more technical debates over how capitalism operates as a system and how marginalized groups of workers are situated within that system. Attempting to move beyond polemical positions that stress the structural significance of class over race or that position racism as a prior condition of feudal production that capitalism progressively supersedes, they add complexity to Marxist accounts of the theory of capitalist value by attending to how original accumulation in the forms of appropriation of Indigenous land, systematization of enslaved labor, and imperial conquest of resources was massively expanded in the Atlantic world with the rise of the industrial economy. Capitalism thus fueled racialization, which was central to its operating logics. Slavery and colonization were not simply premodern or feudal arrangements but were systematically expanded as colonial states developed extractive enterprise linking intensified control of land to intensified control of labor. Although the breadth of the literature on race and capitalism allows for different emphases and lines of argument, as well as for accounts that integrate attention to the significance of gender and sex in the reproduction of colonial capitalism,[6] one key takeaway is a growing emphasis on how racial differentiation is structurally significant to capitalism rather than an abstraction that might be transcended through establishment of more equal economic relations.

I thus use the term "racial capitalism" advisedly to refer to how the development of transnational capitalist networks of production, labor, trade, consumption, energy, and finance structurally relies on the racialization of human national, religious, linguistic, and ethnic differentiation as a condition of systemic expansion or consolidation. In his foundational study of racial capitalism and challenges to it emerging from the Black radical tradition, Cedric Robinson argues that the rise of the capitalist system—first within medieval Europe and later across the Atlantic during colonial expansion of the plantation economy beginning in the sixteenth century—required the consolidation of social identities as groups of merchants and traders attempted to reproduce profits by establishing competitive distance trades in raw materials, agricultural products, manufactured goods, and enslaved peoples. There was an ethnic character to such processes in which trade helped to consolidate geographic identities, for as trade grew and helped to fuel processes of labor migration, industrialization, and urbanization in Europe, at least two forms of human differentiation became embedded in the systemic expansion of capital: (1) the accumulation of labor was regularly a matter of conquering or recruiting subaltern populations whose geographic origins, language, religious practice, or ethnic difference from those who organized industry and trade made them more easily exploitable, and (2) the scaling up of industry and trade helped to generalize the character of human differentiation from localized ethnic groupings to abstracted differences attributable to national and, by the eighteenth century, continental racial groups.[7] Such abstractions of group kinship through race were embedded in the overseas corporate ventures of the colonization of the Americas, which linked the seizure of bonded African labor, the development of European trading networks, and the violent appropriation of Indigenous land.[8] For Robinson, the fundamentals of racial differentiation thus preexisted the rise of capitalism and the development of post-Enlightenment scientific racism. As merchants developed connections between agricultural zones, urbanizing production centers, and seats of monarchal or state authority in the early modern period, Robinson argues, they developed more cohesive commercial identities that scaled up from city to nation-state and, eventually, continent-wide racial designations.[9]

Key to Robinson's argument is the reversal of the assumption that class is the structural basis of social identities across space, while race is an abstracted, epiphenomenal category that arises as a result of class relations; in such a model emphasizing the division of the economic base from cultural superstructure, class forms the universal categories of human social differentiation, while race refers to particular contexts in which laboring classes are pitted against one another in the relations of production. Such a model suggests that the categories of race tend to mask the totalizing economic prerogatives of the globalizing division of labor. In contrast to such an approach to race, the theory of racial capitalism argues that racial differences are reproduced precisely because they are the means for structural expansion and consolidation, including through the connections generated among the control of labor, the colonization of land, militarism, and the centralization of power in the racial state.[10] For Robinson, "the tendency of European civilization through capitalism was thus not to homogenize but to differentiate—to exaggerate regional, subcultural, and dialectical differences into 'racial' ones."[11]

Valuing "Black Gold": Extracting Gulf Oil's Racial Logics

If race was significant to the organization of labor and the accumulation of resources in the development of capitalism, it remained explicit in the labor structures of Gulf oil throughout its history. Consider one of the most notable examples: the racial stratifications in U.S. and British oil firms that controlled Gulf oil in the early and middle decades of the twentieth century. James Terry Duce, at the time an executive of the U.S.-owned Aramco oil company, described to the U.S. State Department the firm's 1949 deportation of Pakistani workers from its Saudi oil fields as necessary, since the workers followed "the Communist line, particularly as regards evils of capitalism and racial discrimination."[12] In *America's Kingdom*, Robert Vitalis examines the history of Aramco's development of a system of racially segregated work camps, separating white U.S. Americans, Italians, South Asians, and Arabs in the Saudi oil fields. Echoing a preexisting segregationist U.S. rhetoric that linked aspirations for civil rights and Black liberation to hidden communist agendas,[13] Duce exhibits how control of race was central to the control of oil in the early phases of U.S.-run development of the world's largest oil reserves in Saudi Arabia.

In retrospect, Duce's statement also anticipates a more recent development: the use of deportation schemes in the 2010s by the Gulf states, most notably Kuwait and Saudi Arabia, to build domestic labor force participation. As part of Saudi Arabia's Nitaqat or "Saudization" labor policy, hundreds of thousands of Indians, Pakistanis, and other Asian workers have been deported in order to make room for companies to meet rising quotas of Saudi workers.[14] This followed public reporting and convictions related to a series of labor actions carried out by Indian migrant workers in the United Arab Emirates' construction sector.[15] A reversal of policies that for decades had used non-Arab migrant populations as a buffer against nationalist and Islamist challenges to authoritarian monarchal states in the Gulf, Saudization targets noncitizen South Asian migrant workers—now seen as potentially rebellious and anti-national—for geographic exclusion. In so doing, the new policy encourages the growing trend of such workers to transit beyond the Gulf to North Africa and Europe. Such policies of national immigrant exclusion are central to historical developments in racial capitalism, which initially exploited and segregated Black workers; according to Mae Ngai, the restriction of immigrants by national origin was a later strategy of white supremacist statecraft apparent in early U.S. immigration law.[16] For Iyko Day, such exclusionary measures build on a larger history of racialized labor management—attempts to control laboring bodies following the economic demand for the presence of those bodies—whether accomplished as domestic segregation and secondary citizenship or international segregation and border control.[17]

The comparison of Aramco and the Saudi state's attempts to expel South Asian workers, however, can only be stretched so far. In the six decades of intervening history, the size of the expatriate workforce in the Gulf—dominated by migrants from a handful of Asian countries whose state policies emphasize remittance economies and human capital export (India, Bangladesh, Pakistan, and the Philippines)—has grown dramatically, reflecting that racially stratified labor generated by oil involves not only the physical extraction of fossil fuel in the oil fields but the manner in which oil profits finance massive development projects, fueling demand for millions of skilled and unskilled contract laborers in such fields as construction, health care, computing, and domestic labor. Although mass immigration is sometimes depicted as part of the "resource curse" that produces various governance problems for major oil-producing countries, the accumulation of migrant labor in Saudi Arabia, Yemen, the United Arab Emirates, Kuwait, Bahrain, and other oil producers in and around the Persian Gulf has long been an explicit strategy that appeared to serve state interests: South Asian workers, especially Muslim ones, were viewed as docile, alienable from family structures, disconnected from nationalist or Islamist political movements in the region, and controllable by work contract. The combination of spatial alienation and lack of citizenship makes immigrant laborers in the Gulf subject to specific forms of control that exacerbate exploitation. Under the kafala system governing contract migration in the Gulf states, foreign workers are required to obtain a contract from a sponsor, who often holds the passport of the worker and commonly exerts control over the worker's movements and schedules. Physical abuse, sexual assault, and backbreaking labor have been widely reported, echoing abuses in earlier systems of colonial contract labor, such as the British and Dutch indenture systems that took place following the end of the Atlantic slave trade.[18]

The ways in which the compromised citizenship status of migrant laborers in the Gulf is exploited is only part of the story of the integration of the Gulf oil economy into the racial circuits of labor and capital on a planetary scale. The basis for the colonial accumulation of oil wealth in the region began with the geological transformation of desert into a site of colonial power politics. Such control was not exerted in a manner that simply reflected a desire by colonial states to accumulate oil, suggested in accounts by both historians and activists on the left in attributing the economic motive of expanding access to cheap oil under U.S. and British adventurism in Iran, Iraq, and surrounding countries.[19] As Timothy Mitchell argues in *Carbon*

Democracy, control of oil's profit potential involved the overriding desire by oil firms (initially U.S., British, and Dutch firms and subsequently state-owned oil companies in Iran and the Gulf) to carve up geographic concessions in order to create vertical monopoly control of underground reserves, thereby *limiting* oil extraction and artificially inflating its price.[20] Thus, in the transformation of desert from a colonial frontier configured as wasteland into a settled extractive enterprise, oil prospecting exhibited a financial logic based on a particular management of the temporality of extraction relative to supply: long-term profit was best realized when large underground oil reserves could be extracted, refined, and distributed at a slow pace that avoided descent into competitive oversupply. The colonial oil oligopoly thus exhibited competition for the control of geographic concession areas while generally militating against open competition to flood international markets. The overriding desire was control of land in a system that would require minimal investment by ensuring limited production.

Such an organization of oil lands involved a kind of franchise colonial model that structured valorization of oil in terms of intricate geographic divisions linked to the emerging structures of colonial states. This carries some important consequences for how we might conceptualize the relationships between oil, colonial geography, and race. Traditionally, settler colonial enterprises that emerged when the British sought to export surplus populations from the British Isles to the white settler colonies (the United States, Canada, Australia, New Zealand, and South Africa) were justified based on a liberal vision that dismissed Indigenous land use as savage and underproductive, whereas settler plantations were understood to more efficiently produce greater commodity surplus, rapidly raising the standard of living at a national scale. Critically reflecting on such colonial notions of progress among liberal political thinkers, geographers Vinay Gidwani and Rajyashree Reddy characterize "waste" as the political other of capitalist "value." Noting that John Locke, John Stuart Mill, and Adam Smith—each of whom articulated colonial notions of political economy and liberal government in the seventeenth and eighteenth centuries—placed significant emphasis on the idea that colonial enterprise was morally justified in appropriating "waste" lands to make rational human use of resources, Gidwani and Reddy characterize the significance of waste as simultaneously "the unenclosed common, the external frontier, and the ethical horizon of civil society" where "the transformation of 'waste'—idle land and nature's bounty—into something useful" comes to define "political modernity."[21] In a key passage of his 1776 book on national wealth, Adam Smith describes settler progress as such: "The colony of a civilised nation which takes possession either of a waste country, or of one so thinly inhabited that the natives easily give place to the new settlers, advances more rapidly to wealth and greatness than any other human society."[22]

If liberal political economy in the contexts of British North American settler colonialism and the franchise operations of the British Raj configured the question of waste around issues of agrarian production—the extent to which land was considered efficiently used to sustain commodity agriculture[23]—the history of colonial oil exploration by British, Dutch, and U.S. corporate interests in the Gulf involved a somewhat different lineage of speculation and corporate rivalry. In the desert, the focus was on mining concessions for oil exports rather than on using the labor of a colonized population to enhance wasted lands. Beginning with Anglo-Persian Oil's discovery of oil in northwestern Persia in 1908, northern oil interests set up a system of concessions for exploration in different areas of the Gulf region that involved negotiation with local leaders. Within this context, the House of Saud began bringing in foreign technical advisors for oil exploration after World War I, when Ibn Saud was one of several local autocrats vying for control of Arabia. Although at the time geologists thought there were little prospects for discovering oil in the region, the major oil companies fought over exploration concessions, and American engineers working in Arabia and Bahrain eventually located oil in the 1930s. In 1938, oil was first struck at the Damman oil fields, which would become the world's largest proven reserves.

Unlike the agricultural basis of North American settler colonialism—which involved a tandem process of land expropriation and labor appropriation that, for Iyko Day following Patrick Wolfe, united native *elimination* (the separation of Indigenous peoples from their land through genocide, removal, and assimilation policies) with labor *exclusion* (the juridical and racial separation of alienated laboring classes imported for the development of colonized land)[24]—the colonial oil franchise depended on a much smaller colonial footprint in order to develop large profits on the international markets. Unlike farmed commodities, there was not a large requirement of labor to apply to the land in order to make it productive. Instead of mixing human and animal labor with land and solar energies as in seasonal agricultural production, oil extraction harnessed deep geological processes that created a flow of energy that could fuel industrial development on a wide scale based on relatively small infrastructures of extraction. Oil enabled a kind of time travel in which the fossilized bodies of long-dead nonhuman species are processed by the earth over the deep time of geological change, then extracted by humans and combusted in the present to speed communication and infrastructural mobilities. Although one might reasonably argue that neoliberal mobility isn't strictly dependent on oil given the role of coal and other energy sources, it is notable that oil is extracted, drilled, stored, stolen, and moved in a patchwork of infrastructures, enabling long-distance ground and oceanic shipping on a planetary scale even as it is embedded in complex forms of statecraft, decentralized militarisms, piracies, and underground forms of speculation and profit.[25] This more-than-human time travel across uneven geographies of profit and environmental destruction enabled by oil allowed both the oil firm itself and the extractive drilling enterprise to mobilize flexibly, even though oil production did require the importation of small reserves of skilled labor to manage drilling, refining, and exporting infrastructures.

What type of colonial racial form did this produce given that low labor input enabled large energy output and thus large profitability? Answering this question involves tracing the relationship of oil to processes of state formation in the region and beyond, with particular attention to how oil was embedded in the transformations of the international order in the interwar period. To begin, the technologies that govern the underground extraction of oil as well as its distribution (through pipelines, large ocean tankers, and so on) involve smaller laboring populations and are relatively closed off to worker control. Mitchell notes that although this makes workers' efforts to exercise power over production more limited than in the case of coal (miners could easily disrupt extraction at the origin), there remain choke points in shipping and pipelines that can be used as sites of collective attempts to wrench control from the oil firm.[26] Regardless of the fact that a certain potential for labor unrest remained, the low relative concentration of labor in the oil sector conformed well with the fact that British and other colonial powers no longer sought to use white settler population expansion as a safety valve for the disposal of convicts and other domestic populations configured as surplus.[27] Nonetheless, there was organized opposition to control of oil by outside powers in the decades leading up to the discovery of large oil fields in Arabia in the 1930s. According to Mitchell, oil firms operating in the Gulf had to contract for concessions in ways that tied their fortunes to the development of autocratic state structures in Iran, Saudi Arabia, Oman, and elsewhere in the region. In the case of Saudi Arabia, the effort to link different regions formerly under Ottoman control by the House of Saud involved alliance with social movements in the region, particularly the conservative *muwahhidun*, whose revivalism involved imposing forms of religious monotheism, abjuring idol worship, and criticizing nomadic social practices of desert tribes. As Ibn Saud allied with this movement, he also entered into contracts with first British and then U.S. oil authorities, whose influence was opposed by the *muwahhidun*. Nonetheless, after crushing the *muwahhidun* uprising in 1927, Ibn Saud was able to strike a deal with them to maintain their support as long as his profits from agreements with Standard Oil could be partly invested in religious education at home and abroad.[28] According to this history, the resulting alliance between U.S. and British colonial oil firms, the emergent Saudi state, and evangelist religious movements in Arabia thus reflected a client patronage system that emerged around oil in the region, tying carbon production to both emergent ethnoreligious identity and the generation of dependent states that purported to represent localized forms of governance and morality.

Once oil operations were embedded under these agreements, oil companies began importing transnational labor forces. Although the initial concessions generally included clauses giving preference to Arab workers, by the 1920s, following the end of the Indian indenture system, British companies especially began to recruit Indian workers to Iraq and Kuwait; by the end of the 1940s, approximately 7,768 workers categorized as "Indian" were employed across Kuwait, Bahrain, Saudi Arabia, Iran, Iraq, and Qatar.[29] Tightly controlling unionization and labor protest, Aramco established a racially stratified labor force in the oil camps, with white, Indian, and Arab workers paid on a declining scale for similar work. Such was still the widespread practice in British and U.S. firms even after the purported shifts in public views of racism and eugenics after World War II. Strikes within the Aramco camps both challenged racially unequal wages and argued for broader democratic reforms to the state, but were later crushed with assistance from the Saudi government. In this sense, the form of the state, the labor system, and maintenance of colonial concessions generated a racialized notion of the relation of sovereignty to oil, with oil profits becoming at least partially subject to the process of articulating Islamic identity with the identity of the state, and those same profits in turn contributing to state control of both political resistance and worker solidarity. As such, Mitchell's argument in *Carbon Democracy* that "racism" could not dictate the organization of colonial oil monopolies because it had become "an embarrassing system of social and political organization" fails to capture the manner in which the reorganization of the former Ottoman-controlled lands of

Arabia in the era of oil discoveries involved the reorganization of ethnic identities and the beginnings of a new racially stratified labor system in the Gulf as part and parcel of oil-funded statecraft.[30]

The Petrodollar as Imperial Currency

From this vantage on West Asia's role in the beginnings of a new racial capitalist circuit of oil production, it is possible to articulate further how oil fueled neocolonial transformations in labor and finance beyond the region itself. Such developments became central to the restructuring of capitalism on a planetary scale in the postwar era. West Asian oil production, South Asian migrant labor, and East Asian manufacturing played conjoined roles in transforming the logics of U.S. power. Before expanding to that story, a caveat about the role of Asia in twentieth- and twenty-first-century capitalism is in order. I endeavor to not invoke orientalist depictions of Asia or Asian economies in a generic form that replicates colonial mappings of the continent as a frontier of capitalist markets, a space of failed or authoritarian states, or a region of geopolitical domination for the United States, Europe, and Russia. Nor do I wish to replicate some accounts of the rise of China in the global economy that have begun to romanticize its displacement of the United States as a world power.[31] It is necessary to recall that "Asia" remains an overburdened signifier, subject to the breadth of colonial mapping and the agendas of area studies research that reproduces it as an arena of difference and intervention. For the purposes of this book, it is necessary to consider how particular Asian states played a pivotal role in the restructuring of the international system after the collapse of the British empire and the rise of U.S. power in the international system following World War II, enabling certain inter-Asian economic connections that are crucial for understanding the transition of colonial capitalism into a highly financialized, dollar-based form dependent on transoceanic shipping of commodities fueled by oil. Thus, specific Asian countries—as oil-producing states, suppliers of labor, client states buttressing U.S. militarism, and developers of carbon-fueled manufacturing and trade logistics—have played key roles in the intertwined neoliberal economic crises and carbon-driven ecological crises of the past half century.

The story of how oil transformed inter-Asian migration pathways, creating the networks and corridors traversed by many of today's so-called climate migrants, begins with the sudden and decisive transfer of petrodollar wealth from the United States, Japan, and Europe to the Gulf states in the early 1970s. On August 15, 1971, Richard Nixon ended the post–World War II Bretton Woods regime of monetary policy when he announced that U.S. dollars would no longer be exchangeable for gold. Nixon claimed this move would "protect the position of the American dollar as a pillar of monetary stability around the world" against foreign speculators. In reality, Nixon's action followed several years of movements among European governments, which had begun a run on U.S. government debt due in large part to massive overexpenditures on the Vietnam War. By 1971, states led by Germany and Switzerland began to pull out of the Bretton Woods framework, floating their currencies and revealing the relative weakness of the dollar. In response, Nixon framed his own withdrawal from the gold standard as a priority to maintain national economic strength and global financial stability:

> Who gains from these crises? Not the workingman, not the investor, not the real producers of wealth. The gainers are the international money speculators. Because they thrive on crises, they help to create them. In recent weeks, the speculators have been waging an all-out war on the American dollar. . . . Let me lay to rest the bugaboo of what is called devaluation. . . . If you are among the overwhelming majority of Americans, who buy American made products in America, your dollar will be worth just as much tomorrow as it is today. The effect of this action in other words will be to stabilize the dollar.[32]

Contradicting Nixon's speculation, a prolonged economic crisis followed, as the currencies of several wealthy countries plunged, inflation rose, and the financial system witnessed prolonged instability in currency and commodity markets. Because dollars were the trading currency for international oil transactions, the depreciation of the dollar sunk revenues for oil producers. In response to Nixon's action, the Organization of the Petroleum Exporting Countries (OPEC) established a policy to fix oil prices to gold. But the real transformation in oil wealth came two years later, when Arab oil producers of the Organization of Arab Petroleum Exporting Countries (OAPEC) announced an oil embargo against the United States and its allies for their support of Israel in the 1973 October War. Over five months, OAPEC exerted export controls and production cuts that quadrupled the price of oil on the world market. The resulting sudden influx of oil dollars to Saudi Arabia, the United Arab Emirates, Kuwait, Libya, and other oil-producing states radically transformed many of the region's economies. Investment by Gulf states increased tenfold over the course of the decade.

In Saudi Arabia, growth leapt to an unprecedented average annual rate of 27.8 percent.[33]

As oil-rich countries experienced unprecedented oil profits and rapid growth in the 1970s, massive state investment in infrastructure created a division among the non-aligned/third world countries. The spike in oil prices and resulting global inflation devastated most formerly colonized states, as their economies—many of which lacked diversification and relied heavily on natural resource extraction—were dependent on oil-fueled foreign imports of basic goods. As the oil boom miraculously sustained U.S. financial and military supremacy in the international system, the World Bank, the International Monetary Fund, and imperial donor states stepped in to "rescue" cash-strapped third world countries with predatory loans that required these countries to abandon welfarist and socialist state projects and accept trade liberalization. As the United States and Europe enforced payment on these debts (in essence using sovereign coercion to force the world's poor to bail out the bad loans of northern banks), many of the poorest countries attempted to diversify from their monoculture dependencies on extractive resource exploitation by growing neocolonial industries such as tourism or by granting concessions of national sovereignty to transnational capital, often through the establishment of tax-exempt export-processing zones and new land arrangements allowing private corporate ownership of extractive enterprise. The subsequent development of the anti-globalization movement—announced with major protests at World Trade Organization, World Bank, and International Monetary Fund meetings in the early 2000s—brought international media attention to the decades-long impacts of these southern debt traps, which effectively precluded social investment and massively increased economic inequality in many of the world's poorest agriculture- and mineral-dependent states. In such locations, there was in the final case increased pressure to export migrants as a source of access to foreign capital remittances and thus to develop the Global South as the prime mover of "human capital" in the world system.

This produced ripple effects globally, as much of the Global South became dependent on oil-fueled shipping, export manufacturing, remittances, and the importation of dollars while the rich countries deindustrialized, moving toward service-centered economies. In the process, Asian nations with large "surplus" laboring populations expanded outmigration. For South Asia, which had already been the source of significant numbers of colonial and "brain drain" migrants in prior eras, there was a turn in outmigration destinations from Europe and the British settler colonies (the United States,

Canada, and Australia) toward the Gulf states, which sought large numbers of Indian, Pakistani, and Bangladeshi workers initially for unskilled construction and industrial labor and later for other categories of domestic and service labor. Not only did the rapid economic expansion of the Gulf states following the oil price spikes of 1973–74 and 1979 generate the largest Asian labor migration corridor in history (eventually expanding across South and Southeast Asia, as I discuss in chapter 3), but it led to the development of production, trade, and financial networks that would sustain United States imperial power through the global economic crisis of the 1970s.

The formation of OPEC and the attempt of oil-rich countries to disrupt the control of oil by the Seven Sisters cartel (Anglo-Iranian Oil, Gulf Oil, Royal Dutch Shell, Standard Oil–California, Standard Oil–New Jersey, Standard Oil–New York, and Texaco—now BP, Chevron, Shell, and Exxon-Mobil) was part of a larger attempt by the decolonizing states of Asia, Africa, and Latin America to develop sovereignty over economies that, due to colonial extraction practices, tended to depend highly on a single commodity. Nonetheless, the coordinated efforts of OPEC were narrowly limited to economic questions and, especially under the influence of Saudi policy, accommodated oil price planning to the broader prerogatives of the United States. Attempts to develop state sovereignty over oil in the Gulf in the 1970s thus failed to develop a broader third world project of cooperation among southern elites, instead exacerbating divisions between oil-rich states and agriculture- or mining-dependent states in Asia, Africa, and Latin America.[34] Yet as Iran and Gulf Arab states increasingly took control of extractive enterprises, oil became a potent commodity for state-led development schemes and an important tool for forging third world alliances and frictions.

If the oil boycott was ostensibly an effort by majority-Arab states to counter U.S.-funded Israeli settler colonialism in and beyond Palestine, the actual effects of the policy occurred broadly in the international system rather than affecting U.S. policy in Israel. First, by asserting pan-Arab sovereignty over oil in a narrowly economic fashion, the oil producers created a division among third world states that had been agitating in the United Nations and the Non-Aligned Movement for coordinated action to ensure profitable prices on key natural resource exports. Instead, oil exporters engaged in a much narrower action that massively increased their own extraction profits at the expense of the many oil-importing southern countries. Prospects for a leftist alliance among southern elites to challenge some of the neocolonial inequalities of the international economic order have never recovered, and even OPEC has persistently struggled to maintain discipline on production quotas among its own members, threatening to unravel the monopolistic exploitation of oil profits established earlier by the concession system of the colonial oil firms.[35] Second, despite producing a supply crisis that briefly affected the U.S. consumer market, the recycling of Gulf petrodollars back to the United States (as Gulf states moved profits to U.S. banks and began importing more goods, including arms) ultimately helped to sustain the dollar and to secure the central place of the United States in international finance. There is no indication that the end of the dollar standard was a strategic financial strategy of the Nixon administration to maintain U.S. hegemony after Vietnam; however, the oil shock allowed the United States to reconsolidate a central role in international finance based on the establishment of the dollar as reserve currency for international trade. The limitations of a state-centered, elite version of national decolonization practiced in the Non-Aligned Movement were made clear in this moment of crisis. The outcome of these moves by the United States and the Gulf oil producers created a massive transfer of wealth to the Gulf, which would in turn buttress use of the dollar as the world's international reserve currency—a development that breathed life back into U.S. finance even as the U.S. manufacturing base shrunk and U.S. monetary policy weathered a series of crises around diminishing exports and inflation.[36] Despite the role Vietnam War debt played in helping to precipitate the global economic crisis of the 1970s, U.S. government debts could continue to be sustained due to the maintenance of the dollar as reserve currency. This debt exceptionalism—which finances the massive U.S. military footprint and other deficit spending—continues into the present, as the United States remains the only country whose balance of payment deficits does not substantially affect its currency or capacity for financing.[37]

These transformations in finance wrought by oil came at a time when the United States and much of Europe were deindustrializing, manufacturing bases were shrinking, and wartime military Keynesianism was being supplanted by privatization, massively shrinking working-class employment. The impact on Black and Latinx workers in the United States was acutely devastating, as they were less insulated by seniority and increasingly subject to unacknowledged forms of employment discrimination. As Asian manufacturing expanded during this contraction for Europe and the United States, massive waves of rural-to-urban migration accompanied the new industry. Although xenophobic discourses about job loss tend to suggest that Asians took jobs from whites, the human costs of adaptation to transnational regimes of production were massive, as state attempts to organize production to enhance capitalist development involved land dispossession and large-scale migration.[38] If the oil shock led to large development projects in the Gulf that required new sources of labor, it also accelerated the reorganization of manufacturing accelerating new migration pathways from rural to urban areas across Asia. Although on a world scale, this coincided with the feminization of industrial labor, seen as more flexible by states and corporations alike, in the Gulf region, such migration followed a longer historical pattern in which male migrants departed first and women gained a small but growing share of the flows over time, particularly as demand for domestic and care laborers grew.[39]

Finally, much of the petrodollar surplus went to banks in the North, bailing them out of unsustainable debts and reinforcing a shift toward predatory lending to southern countries hobbled by inflation. Overall, as Gulf states received the largest postindependence transfer of wealth across borders,[40] non-oil-producing states widely experienced recession and were ultimately forced into debt traps under the neoliberal conditions of the Washington Consensus, which erroneously suggested that "globalization" lifted all boats, including the global poor. Southern states (including Bangladesh) that accepted loans from outside donors were required to shrink the size of the state and to accept trade liberalization. Thus the dream of realizing national sovereignty over oil among southern elites helped contribute to the nightmare of neoliberalism's entrenchment of economic inequalities and dependencies across borders. Not only did the Arab oil boycott sustain U.S. empire through its deepest financial crisis, but it helped to hobble the socialist or welfarist ambitions of countless nations attempting to redress the domestic and international inequalities created by colonialism, making them more dependent on market liberalization and export-oriented development.

For these reasons, some commentators argue that "empire" is no longer the proper term to describe the inequalities of the international system; a system of rentier finance capitalism may deposit profits in highly unequal ways across the Global South rather than concentrating power only in the North.[41] But it is important to not jump too quickly to view the rise of the Gulf petrostates or East Asian industrialized economies as usurping U.S. empire, given their deep dependencies on U.S. consumer markets and dollar-denominated finance. While small ruling minorities win much of the economic gains due to oil profits, manufacturing profits, and speculative finance in Asia, urbanization and migration processes fuel forms of ethnic differentiation and conflict as well as continued precarity for those engaged in subsistence agriculture and small enterprise in the rural peripheries. Meanwhile, the maintenance of both dollar supremacy and the massive U.S. military footprint by cycling oil and dollars through the Gulf states helps to maintain an exceptional position for the United States in the geopolitical order. Although the dependence of several of the Gulf oil producers on the United States for rentier growth has come under duress since 9/11 and the expansion of unconventional oil production to diversify U.S. imports, the shifting of the oil supply has not significantly undercut the global demand for Gulf oil among a variety of importing states. One outgrowth of this situation is that regional powers led by Saudi Arabia—now the world's largest arms importer—use their dollar reserves to purchase weapons, fueling the arms race in the region and contributing to the entrenchment of conflict within and between states.[42] This has been most notable in Saudi Arabia's prosecution of a war and humanitarian disaster in Yemen, but it has also been a longer-term trend in the region as U.S. and Russian interventions, Arab Spring protests, and regional conflict spur monarchal states to securitize against regional and global powers.

As such, the consolidation of the dollar as the world's reserve currency following the 1973 oil embargo may not exactly constitute a preplanned strategy of empire, but it does present a case in which U.S. officials and Gulf states worked in concert in ways that used oil to buttress U.S. power in the world system, which resulted in massive forms of war-driven displacement and racialized contests over migration in the present. This represents a de facto continuation of U.S. empire through the crisis of the 1970s by establishing financial dependencies between the United States and rising Asian economies.[43] This power did not mainly take the geographic form of direct colonial domination of different Asian nations by the United States but rather emerged from the financial implications of oil's establishment of the dollar as the world's reserve currency, which allowed massive borrowing for military ventures, the expansion of arms importation for cash-rich Gulf states, and the development of Indian Ocean and Pacific shipping routes, which enabled the rise of East Asian manufacturing. This last point became crucial as oil shipments from the Gulf to East Asian states, along with domestic increases in coal consumption, helped to fuel the transnational system of manufacturing for products ranging from small electronics to automobiles, textiles, and a variety of consumer goods. In the process, migration became a key adaptation strategy for several Asian nations facing the overlapping economic shifts of oil-driven inflation, population growth, structural adjustment, and, in some cases, rapid currency devaluation.

Thus, the use of oil extraction to create a path for oil-rich countries to speed development in the 1970s enhanced economic divisions between the non-aligned countries as South and West Asia were incorporated into oil-fueled circuits of globalizing trade.

Debt, Development, and Displacement

Robinson's account of racial capitalism was attentive to shifts in the scale of migration, moving from the accumulation of labor across the Mediterranean in southern Europe during the rise of capitalism to the mass kidnapping and commoditization of enslaved labor in the Atlantic trade. Robinson establishes that spatial alienation is one of the key forces driving the systemic reproduction of racial differentiation and management as capital shifts human populations that come into contact through circuits of exchange and migration. Moving beyond Robinson's contexts, we witness increases in the total number of international migrants during the neoliberal era, coincident with the spread of industry, the reorganization of agrarian lands under large enterprise, and the establishment of export-oriented manufacturing. New waves of migration from South to North emerged following the end of World War II, following changes in citizenship rules in the colonial powers and the economic transformations of national independence in many areas of Asia and Africa between the 1940s and 1980. In the neoliberal historical juncture since the oil shock, there was an increase in both transborder and internal migration as basic features of economic life in many countries, compounded by the intensifying wealth inequalities, resource scarcity, and immigration restrictions in the North. Based on UN estimates, international migration increased from 79 million annual migrants in 1960 to 250 million in 2015, by which time approximately 60 million of those were defined as refugees. The two major spikes occurred in 1989 and 2015, with the fall of the Berlin Wall and the Syrian war generating mass displacement. Although in sheer numbers the United States and Germany currently remain the receiving countries with the largest population of international migrants, there has been marked growth in South-South migration, and Saudi Arabia and the United Arab Emirates are numbers three and six, respectively, on the list of top receiving countries. Other countries around West Asia—including Iraq and Lebanon—have experienced major migration flows as a result of post-9/11 wars.[44] Approximately half of the current transnational migrants originate in Asia.[45] One of the main shifts that happens with larger-scale migration, especially migration of agrarian populations, is urbanization and the creation of multiethnic cities. Whereas in 1950, 30 percent of the world's population was concentrated in cities, by 2008 this number had surpassed 50 percent. In China alone, the boom in manufacturing and urban-led development created what is likely the largest and fastest urbanization process in history, with the urban proportion of the population rising from 36 percent in 2000 to 50 percent in 2010 (comprising some 670 million people).[46] In South Asia, extreme poverty remains a key driving factor for rural-to-urban migration, and this migration tends toward the feminization of the labor force, which is preferred by transnational industry.[47] Because the proportion of the population dwelling in cities remains the lowest in Asia and Africa, these are the regions experiencing mass agrarian displacement and the most rapid urbanization trends with the fastest-growing cities.[48] The rise in various forms of xenophobic targeting of Asian and African immigrants outside their countries of origin has become commonplace, most spectacularly witnessed in the racist media reporting during the so-called European migrant crisis of 2015.[49] Population growth remains an extremely unequal phenomenon, as speculation in land and urban housing continues unabated while poor migrants increasingly settle in large shantytowns at the urban periphery.[50] As in other parts of the world, it is common for states in these regions to establish policies to discourage continued rural-to-urban migration to prevent overly rapid population growth in cities.[51]

It does appear that environmental pressures generated by the oil economy contributed to migration pressures. With sea level rise, extreme heat, particulate pollution, and increased flooding affecting large population centers across Asia, the vulnerability to weather-driven displacement and environmental toxicity has had a measurable impact on premature death rates and crop failure.[52] Although coal may have played an outsize role in historical carbon emissions, oil's role in the era of neoliberalism rapidly expanded emissions and, more significantly, tied those emissions to the systemic production of growth by creating the mobilities required for distance trades. One estimate suggests that two-thirds of carbon emissions over the last two centuries have been generated by the oil complex.[53] So although coal is more readily substitutable by renewables in national energy infrastructures, oil is more deeply embedded across different technologies, including in shipping and other critical arenas of logistics that fuel transnationalization. Oil systemically reproduces dependence on itself, energizing circuits of migration through both the inequities and the environmental disasters it generates, particularly for coastal and agrarian populations.

In retrospect, the conjoined economic and environmental costs of neoliberalism were often minimized in English-language media focusing on globalization in the 1990s. Failing to account for the uneven geographies of accumulation and dispossession that came with the mass population movements and urban transformations of neoliberal economic globalization, U.S. and European journalists and economists writing after the fall of the Soviet Union often characterized globalization as benign or positive in its effects on human development. Such pronouncements worked in concert with triumphalist declarations of "the end of history" that began to frame the new postsocialist era for both the former Eastern bloc and nominally socialist states in Asia.[54] Such expressions of the free-market ideology ignored at least four key facets of the neoliberal transformation that are important for understanding migration in the age of climate crisis: (1) the expanded economic dependency and deprivation wrought by the debt crisis affecting poorer countries, (2) the use of migration as a solution to the mass pauperization of the poor, (3) the externalization of the violent effects of amplified fossil fuel consumption into the biosphere, and (4) the centering of oil in international finance. Carbon-fueled forms of capitalist networking were pivotal in producing the myths of global integration and modernity even as growing structural racism suggested the underside to this form of uneven development. For Ruth Wilson Gilmore, given "the deepening divide between the hyper-mobile and the friction-fixed" across the international division of labor, globalization must be understood as a phenomenon in which "racism fatally articulates with other power-difference couplings such that its effects can be amplified beyond a place even if its structures remain particular and local."[55] This would be true even for populations that migrate due to political, economic, or environmental pressures—those that experience severe economic and mobility constraint after being spatially alienated from kin networks and resources at home.

The rise of the anti-corporate globalization movement as a challenge to neoliberal structural adjustment by 1999, the worldwide mobilizations against the U.S. invasion of Iraq in 2003, and a series of movements challenging market authoritarians in the 2011 Arab uprisings and their aftermath have clarified the error of Thomas Friedman's assessment that globalization had rendered the world "flat."[56] If worldwide and national levels of economic inequality have soared to record highs during this period as a result of rabid speculation and rentier capitalism, neither neoclassical nor Marxist accounts of financialization since the 1970s have specified the racial logic of this political economy. There is no post-Fordist capitalism (or capitalism in general) that is not a racial capitalism. In contrast to some Marxist analyses that define neoliberalism primarily as an austerity-based financial logic coincident with the shrinking role of the state and of fixed assets in the calculus of value, theorists of racial capitalism today assert that "accumulation by dispossession" is foundational to emergent logics of finance.[57] A key element of these logics has been the proliferation of debt as a manner of securitizing the militarized extraction of labor and land. Irreducible to processes of elimination and exclusion aimed respectively at expropriating native land and accumulating migrant labor, finance's racial logic involves corralling commodities, labor, and assets into ever more speculative and securitized instruments to promote circulation. Because the logic of financialization tends toward expansion of debt and proliferation of speculative bubbles that recast the value and profit capacity for different commodities, fixed assets, and enterprises, it has been transformative in intensifying the effects of enclosures and other racial regimes of property that displace poor people, rendering growing populations as surplus.[58]

Thus, in their reflection on the U.S. subprime housing crisis as an outgrowth of a racial geopolitics of debt, Denise Ferreira da Silva and Paula Chakravartty pose the problem of racial capitalism in a debt-fueled era of financialization as follows:

> How could the predatory targeting of economically dispossessed communities and the subsequent bailout of the nation's largest investment banks, instantly and volubly, be recast as a problem caused by the racial other ("illegal immigrants" and "state-dependent minorities")? . . . Our interest is in situating the racial moment of the financial crisis in the last three decades of neoliberal backlash waged across the postcolonial (global) South. As a starting point for our discussion we assume that these recent histories are themselves embedded in the colonial and racial matrix of capitalist accumulation of land (conquest and settlement), exploitation of labor (slavery, indentured labor, forced migration), appropriation of resources, and ultimately the very meaning of debt.[59]

In this statement, da Silva and Chakravartty build on Robinson's earlier account of racial capitalism. For Robinson, "The tendency of European civilization through capitalism was thus not to homogenize but to differentiate—to exaggerate regional, subcultural, and dialectical differences into 'racial' ones."[60] Given that da Silva and Chakravartty figure racial capitalism as a process that attributes moral failure not just to *persons* but also to *places*, crises of environmental extraction might also be understood as processes that transmute logics of space into time: places where migrants are concentrated, especially in coastal and other climate-affected regions, might appear as falling behind in development and in need of interventions that bring them up to speed with carbon-fueled production and urbanization processes. Even while oil maintains its high price on the international market, buoyed by continuing transborder conflict in the Gulf, the literature on environmental racism shows how zones of environmental precarity and toxicity concentrate the violences of dispossession and debt. Discussions of the intersection of environmental racism and racial capitalism have focused recently on the example of Flint, Michigan, where the loss of Black union jobs and economic decline cleared the way for a revanchist political formation to take control of the city's utilities, eventually leading to the mass lead poisoning of the majority-Black population.[61]

This is also evident in the ways that climate adaptation discourses consider threatened agrarian livelihoods as disappearing sites where migration itself is supposedly a needed development strategy. For example, Kasia Paprocki documents how the embrace of shrimp aquaculture as a development strategy in the Khulna district of Bangladesh's southern coastal zone reflects both that banks and NGOs have embraced large landholders' desire to remove agrarian populations and that they have begun a process of speculating that urbanization will become a solution to this displacement. Naming this process of purported climate adaptation as "anticipatory ruination," Paprocki emphasizes that organizations like the World Bank latch on to such schemes based on long-standing assumptions that urbanization is the path to development and that "development-induced migration should be seen as a (now-necessary) opportunity, instead of a threat."[62] The temporal management of environmental crisis results in a situation in which migration is configured, in advance, as a pathway to national development, reinforcing existing demands to displace agrarian populations and to expropriate land in the name of modernization. In chapter 3, we will witness how this plays out in the Bangladeshi state's promotion of climate refugee discourse as a way to potentially bolster emigration-driven remittance growth.

Toward a Green Capitalism?

If contemporary racial capitalism thrives on speculation of future value, why doesn't the system correct itself and more properly value oil as a declining asset, given the push for renewable energy and the likely scenario of technological transitions away from fossil fuels? The very public push by the Saudi crown for a high valuation of Saudi Aramco's initial public offering gives a glimpse of why oil continued to command a high price even after the onset of the COVID-19 pandemic. After a lengthy campaign seeking a $2 trillion valuation for Aramco in public trading, financial analysts remained skeptical, and the Saudis were unable to launch the IPO on their terms on one of the major stock exchanges. The situation was further complicated by backlash against the Saudi government's assassination of journalist Jamal Khashoggi as well as heightened tensions with Iran, and the Saudis decided to offer Aramco shares on their own national exchange with an initial valuation of $1.7 trillion. Within weeks of trading, however, the stock price rose over 10 percent and topped $2 trillion, buoyed by investors from around the Gulf region. Leveraging the world's largest publicly traded company, the monarchy generated over $25 billion in profit by selling just 1.5 percent of its shares. The purported goal was for the state to invest in a number of economic initiatives to diversify the economy away from fossil fuels.[63] Alongside the Nitaqat policy, this step could point toward a future in which Saudi Arabia moves away from oil and migration dependency and toward new forms of national development. However, claims about how the Saudis will spend their oil profits remain open to evaluation as the kingdom continues to place massive investments in war.

The relatively high price commanded for oil—which is itself buoyed by the petrodollar financial circuit and the regional arms race—reflects the continued manner in which growth, development, militarism, and international trade are closely correlated to fossil fuel use in the present. This is not only so for rich oil producers in the Gulf; it remains the case for poor oil-importing, export-oriented neoliberal states in Asia. As such, it is not just investors who maintain that oil is central to contemporary global capitalism but also international development experts, who have attempted in recent years to develop more environment-based models for wealth measurement.

If oil was one of the primary drivers of a new, financialized regime of inequalities in the U.S.-dominated world system of neoliberalism, intergovernmental and nonprofit development agencies, alongside U.S. and European aid programs, have been the primary forces generating public responses to these inequalities. International development discourses have long framed poor countries as failing to achieve the benefits of modern capitalism because of governance and education failures. Despite the lengthy history in which colonial extraction worked to create colonized states as agrarian or extractive peripheries, and despite the way that the oil shock and neoliberal policies have in the past half century exacerbated the economic dependency of poor countries on outside capital, this framing of southern countries as outside the time of modern capital remains; contemporary racial capitalism is spatially capacious, but its logics are deeply invested in managing the time of the purportedly underdeveloped nations. Although Frantz Fanon, in "The Trials and Tribulations of National Consciousness," long ago warned of the neocolonial economic dependency that emerges from the relation of the newly independent state to transnational capital,[64] development discourses continue to pathologize southern states and work to reinforce conditions in which internal migration (from rural to urban areas) and external migration (to wealthier countries in search of remittances) remain a key pathway to achieving a higher degree of "development." In fact, attempts to integrate environmental factors and attention to climate-driven migration into such calculations of underdevelopment may in fact deepen international wealth inequalities by attributing speculative value to southern environmental systems—value that remains unrealized as actual capital to the agrarian poor or even to the neoliberal state. Paradoxically, the message of national wealth estimates based on natural capital is not that natural capital valuation will help these countries hold natural capital in reserve to prevent further environmental harm but that these countries must transform surplus natural capital in order to remedy deficits in produced capital—especially infrastructure and human capital—which may include exportable migrant labor. Thus, in working to conform the space of climate-affected regions to the temporal management of development as a national process of obtaining wealth, neoliberal development discourses and strategies continue to reinforce migration as they incorporate new thinking about environmental problems.

In the 2018 report *The Changing Wealth of Nations: Building a Sustainable Future*, a group of economists writing for the World Bank describes how capitalist development depletes the environment of a country. In this process, natural resources are converted into either produced or human capital. In their narrative, the economists develop a parallel concept of natural capital, which involves estimating the wealth potential of environmental systems in a particular country. Notably, in the description that appends natural capital to a developmental narrative about increased national wealth, oil-producing countries in the Gulf region become an exception, since they are able to attain high-income status without depleting the resource base:

> The share of natural capital gradually declines as countries graduate from low- to middle- and high-income status. Human capital reaches 70 percent of wealth in high income OECD countries (and natural capital only 3 percent)—not by reducing the amount of natural capital but by adding more produced capital, especially human capital. This makes sense because economies can only move beyond subsistence production of food and shelter to manufacturing and services with the addition of human capital, infrastructure, and other produced capital. The exception is high-income non-OECD countries, dominated by the high-income oil and gas producers of the Middle East, where natural capital remains a large component of wealth.[65]

The Gulf oil economy here remains an exceptional case: it is the one site of extractive enterprise that remains profitable enough that entire national economies with levels of per-capita wealth greater than those of the United States and Europe can be sustained with a primary dependence on the "natural capital" of carbon resources rather than labor, euphemized as "human capital." Outside these contexts of massive oil supply, natural capital is an object primed for conversion to human capital. And to the extent that poor countries fail to achieve development, their failure to effectively convert natural capital into sustainable produced or human capital reflects governance failure. Confirming long-standing criticisms from the anti-globalization movement, such forms of wealth measurement—which could overvalue the wealth of poor nations based on unrealized capital of the environment, justifying lower aid levels—reinforce the existing privatization and extraction models that the World Bank and the International Monetary Fund have long used to hobble attempts at socialist or welfarist development across the Global South.

Reinforcing the vision of natural capital–intensive states as unproductive and outside the time of modernity, such wealth calculation strategies fail to challenge the hegemony of the petrodollar and sustain the international division of debt access. Meanwhile, the temporal management of race—presuming peripheral "wastelands" subject to climate disaster as ripe for redevelopment without their current populations—proceeds apace despite environmentalist efforts to reorient capitalist views of time based on the deep time of environmental change. Studies coming out of traditions of political ecology and green Marxism have placed the crises of capitalism within ecological systems, refiguring the role of what Marx called primary or primitive accumulation within broader relations of extraction and systemic control of planetary metabolic processes. Thus for Jason Moore, the society–nature distinction was one that instituted a progressive narrative of civilization, rendering some humans as relegated to nature against the prerogatives of development.[66]

And yet even as we grant that environmental violences require us to pursue forms of thought and politics that look beyond the human, it should come as no surprise that projects to transcend anthropocentric ethical and political frameworks present new horizons of accumulation. This is an argument that critics of ecotourism and biotechnology have been making for some time. Over the past decade, neoliberal economists—especially those working in the development field—have developed detailed instruments and analyses for thinking natural and economic processes in tandem. Attempts to quantify natural capital—such as the university-NGO partnership the Natural Capital Project, the UN's System of Environmental-Economic Accounting, and the OECD's Green Growth indicators—are in the process of developing increasingly complex models for valuing natural resources not just as tangible assets but also as agents of productivity. Yet the recognition of the economic force of natural resources and processes does not in itself involve an assumption of economic agency. Indeed, they may reinforce the racialization of vulnerable environments—and of development itself—for poor people increasingly affected by climate change and other environmental disasters. The lesson is not that these populations are emerging as new climate migrants but that the very forms of knowledge used to narrate their diminishing present and unlivable future entrench the suppositions of underdevelopment that make the long histories of extractive colonialism and oil-fueled neoliberalism generative of mass environmental racism.

CHAPTER THREE

From Insecurity to Adaptation

Bangladesh, Human Capital, and the Figure of the Climate Refugee

Emerging discourse on climate-induced migration in South Asia demonstrates certain limits of security thinking for adequately grasping the geopolitics and racialized public imagery of climate change. Bangladesh has become the epicenter of Western security thinking about migration for the region (and increasingly the world), as it is the country described as the most prone to climate-driven displacement. Eighty percent of Bangladesh's landmass is floodplain, and the government has begun planning for more regular cyclones and river floods that threaten displacement and mass casualties. Major publications around the world regularly feature images of Bangladeshis displaced by flooding, cyclones, or sea level rise, alternately focusing on stories of personal tragedy and resilience. Resilience, it would appear, is in high demand, as the coming displacements portend regional or even global security crises. Bangladeshi Major General A. N. M. Muniruzzaman (Ret.) of the Global Military Advisory Council on Climate Change regularly speculates in the international press that tens of millions of Bangladeshis may soon be displaced due to climate change, arguing for changes in policies worldwide to admit larger numbers of environmentally affected migrants from South Asia. Speculations of security risk produced by displacement or resource scarcity go beyond the borders of Bangladesh. The apocalyptic climate-driven conflict and migration scenarios narrated by security experts include a potentially nuclear water war between India and Pakistan due to the melting of the Himalayan ice pack; the floodplain displacement of one-fifth or more of Bangladesh's population, creating an international migration crisis; and a Hindu-Muslim conflict caused by a spillover of Muslim climate refugees from Bangladesh into the Indian states of Assam and West Bengal. Speculations of climate-driven conflict regularly invoke racialized geopolitical mappings of Muslim climate migrants as a threat to international order. At the same time, journalists and policy makers increasingly look to areas of environmental change like Bangladesh for lessons about how people might adapt to a warming world. In the process, the climate migrant can at once be configured as a racialized threat to the global order and as the solution, whose environmental knowledge may be appropriated by states looking for ways to manage climate change's security crises.

Although mass flooding and other climate-induced disasters do indeed influence migration and access to resources throughout South Asia, the figure of the climate migrant or climate refugee in South Asia is often depicted as an exceptional and new disruptive force threatening to produce failed states. In the process, speculations of sudden climate-driven state collapse, ethnic conflict, mass migration, and adaptation contrast with some basic facts about the deeply ingrained forms of human mobility in the region: (1) flooding-induced migration is long established in the areas of India and Bangladesh that border the Bay of Bengal, as cyclical river flooding and cyclones have produced temporary displacement and permanent migration before climate change increased such risks; (2) new climate adaptation projects are encouraging migration by reorganizing agrarian production in advance of major coastal inundation; and (3) individuals migrating away from flood-prone areas of Bangladesh continue to follow migration routes established in the 1970s, when neoliberal transformations of the state made rural-to-urban migration to Dhaka and contract migration to the Persian Gulf states the primary paths for those facing displacement, debt, or unemployment. These factors point to one common difficulty in discussing how environment affects migration: it is often difficult to disentangle the environmental influences to migrate from political, economic, and social ones, including pressures to migrate produced by climate adaptation discourse itself. In South Asia today, it appears that environmental drivers of migration overlap in dynamic ways with longer histories of displacement and migration networking. Apocalyptic scenarios of resource wars and mass migration tend to sideline more specific investigation of how extant migration patterns have been incorporating those affected by or anticipating environmental disasters.

This chapter attempts to move away from the exceptionalism, crisis thinking, and geographic generalizations of security narratives on South Asian environmental migration to address the interlinked ways that the expansion of the Gulf oil economy, the U.S.-led transformations in neoliberal systems of finance and trade, climate adaptation schemes, and the waste effects generated by oil consumption are converging to produce contemporary circuits of transnational migration within and from contemporary Bangladesh. By focusing in particular on how the oil-producing Gulf states have played a role in Bangladesh's postindependence development policy, I argue that the migration effects of the carbon economy must be understood not only in terms of the *waste* effects and systemic ecological changes wrought by carbon emissions but also through the *extraction* and *production* processes that have generated transnational growth and exchange, as well as by the shifting development economies that integrate visions of ecological collapse. In the process, it is possible to see how the neoliberal conditions of industrialization and development have generated particular configurations of Bangladeshi migration that are irreducible to the weather. Instead, the emergence of the figure of the climate migrant in Bangladesh may work to hide from view a longer history of neoliberalism as a phase of racial capitalism, wherein oil-fueled inter-Asian trade, manufacturing, development, and migration networks are fundamental to the reproduction of surplus labor in a warming world.

Bangladesh and the Gulf Labor Corridor

The oil boom and the resulting transformations of international finance and trade have resulted in significant shifts in the geographies of the South Asian diaspora. As India, Pakistan, and Bangladesh faced overlapping financial, political, agricultural, and labor pressures during the crises of the 1970s, they joined other Asian nations—including China, Indonesia, and the Philippines—in transforming human capital into a major export and source of foreign-dollar remittances. Much of this outmigration went to the Persian Gulf states, as labor migration became pivotal to completion of the massive infrastructure and development projects there, undertaken with oil revenues. Gulf states began to grow the number of temporary work visas granted to South Asian migrants immediately after the 1973–74 oil crisis. Echoing the colonial "coolie" migrations organized under the British indenture system,[1] the kafala contract labor system established by the Gulf states relies on temporary labor contracts that attempt to prevent migrants from establishing wealth and rights in the destination country. Despite reports of widespread labor abuses and exploitation of migrants in this system, as well as restricted freedom of movement, as employers typically hold migrants' passports, the number of migrants has swelled to such large numbers that immigrant laborers—mainly from countries in South and Southeast Asia—make up 50 to 90 percent of the workforce in Gulf countries.[2] This reflects steady growth in South Asian migration to the region, with an estimated 9.5 million South Asians residing there today, a number higher than the combined estimated populations of those of South Asian origin in the United States, Canada, the UK, and Australia.[3] By the early 1980s, there had already been a decisive shift in migration destination from the brain drain receiving states of the UK and its settler colonies (the United States, Canada, New Zealand, Australia) toward the Gulf.[4]

Three points of historical transition are important for witnessing the racialized character of this migration corridor from South Asia to the Gulf states. First, in the early years of the oil boom, majority-Arab countries remained important source countries for Gulf labor. However, as recruiting agencies formalized processes for importing labor, South Asian countries established laws regularizing contracts for outmigration, and as the private employment sector grew in the Gulf, South Asian laborers were increasingly seen as preferable to Arab ones for three reasons: (1) they could be employed with lower wages and workplace protections, (2) they migrated as individuals rather than in nuclear family units, and (3) they were viewed as less likely to politically threaten monarchal state systems by aligning with social movements.[5] Second, some Gulf states decisively shifted from Arab to South Asian migration following the 1991 Second Gulf War, as Saudi Arabia and Kuwait viewed Iraqis, Jordanians, Palestinians, Yemenis, and Sudanese as untrustworthy, since their home states supported Saddam Hussein's government against the Saudi-U.S. alliance. Over 1.5 million Arabs, including Palestinian refugees who had lived in the region for generations, were expelled from the Gulf, mainly from Saudi Arabia and Kuwait.[6] Third, a second oil price spike in the early 2000s led to massive labor recruiting along the established corridor, bringing in millions of South Asians and an increasing proportion from the smaller nations, especially Bangladesh and Sri Lanka. From 2001 to 2012, the annual number of South Asian migrants entering the Gulf increased from 232,668 to 722,139.[7] These trends reflect how South Asian labor was imported as a racially differentiated population whose status as temporary guest workers made them a solution to political disputes across the greater Persian Gulf region, exacerbated by U.S. intervention.

Bangladesh's first comprehensive emigration law, the 1982 Emigration Ordinance, was aimed at formalizing the growing population of emigrants departing for the Persian Gulf states. Although India provided the majority of South Asian migrants to the Gulf in the 1970s, the establishment of diplomatic relations between Bangladesh and Gulf states in 1975 contributed to the international diversification of South Asian origin countries. By the early 1980s, recruiters were increasingly targeting smaller South Asian nations, including Bangladesh, Nepal, and Sri Lanka. In 1976, there were 6,087 Bangladeshi workers in the Gulf; but by 1979, there were 24,485. During these same years, annual remittances to Bangladesh from the region increased to over US$15 million.[8] But it was largely after the year 2000 that Bangladeshi emigration to the Gulf came to rival Indian and Pakistani numbers. It was at this time that flows of emigrants from Bangladesh moved decisively away from India, the historic top destination country, toward the Gulf. Three-fourths of the outmigration from Bangladesh from 1976 to 2012 went to the Gulf, producing annual remittances of up to US$14 billion; 75 to 80 percent of these migrants went to Saudi Arabia or the United Arab Emirates.[9]

With the promulgation of the 1982 law, labor recruiters were able to bring larger numbers of Bangladeshis to Saudi Arabia, the United Arab Emirates, Oman, Qatar, Kuwait, and Libya as oil dollars rapidly expanded state development projects and expanded markets for unskilled migrant labor. This coincided with the privatization of many sectors of the Bangladesh economy; under both the Zia and Ershad governments in the 1980s, guided by conditions on foreign aid, Bangladeshi policy moved away from state-run economic activity to privatized markets. Despite the fact that the emigration law claimed to curtail abuses by recruiters, lax enforcement of labor standards and the difficulty of filing claims expanded the ability of Gulf firms to recruit under false promises about pay and working conditions.[10]

Outmigration and export-led development have been the twin strategies for gaining access to foreign capital, as the Bangladeshi state incorporated neoliberal structural adjustment. Although in the early years following independence, Bangladesh largely remained in line with the socialist policies of other South Asian states, the Zia (1975) and Ershad (1982) governments increasingly embraced Washington Consensus prescriptions for encouraging development through foreign investment, in the process shrinking the size of the state and expanding private market activity accessible to transnational capital. Both external pressures (foreign aid conditions imposed by the United States and international creditors) and internal pressures to form state coalitions with the military and industry led these governments to advance neoliberal policies, setting up Dhaka as a manufacturing center and expanding state dependence on remittances of emigrant workers. As Bangladesh became a hub for global textile production, migration to Dhaka from rural areas—including many coastal regions subject to flooding—skyrocketed.

In the midst of the largest wave of South Asian migration to the Gulf region, governments became nervous about concentrating demographically similar laboring groups and made a concerted effort to move some recruiting to Southeast Asia, with the Philippines providing the largest emerging labor source. By the early 2010s, the deep dependence on South Asian labor combined with a growing awareness and coordination among workers led to the first work stoppages and protests by South Asian workers contesting withheld wages and poor working conditions. Gulf states responded harshly, expelling workers and developing new policies to "localize" the labor force by recruiting more nationals into the private sector. In 2012, Kuwait expelled seven hundred thousand South Asian workers; Saudi Arabia established its Nitaqat labor law, which ranks corporations based on their proportion of citizen workers and requires that each business employ a minimum 20 percent Saudi workforce.[11] Across the Gulf, states cracked down on the sale of visa sponsorships, deporting hundreds of thousands of South Asian workers. Finally, from 2017 to 2019, Saudi Arabia imposed additional restrictions and fees on migrants, leading to an exodus of approximately one million workers.[12]

Despite these emerging trends that work to shrink the Gulf Labor corridor, the establishment of large networks of family and peer recruitment in both sending and receiving states have helped to maintain and grow large flows of migrants. For Bangladesh, outmigration has continued despite the crackdowns, in that contractors are able to switch destination countries in response to conditions on the ground. With the attempt to clamp down on migration from South Asia to Saudi Arabia, it appears that Bangladeshi workers are shifting to other countries in the Gulf and the Maghreb. With a large labor recruitment network already in place, it is possible for more Bangladeshi workers to enter neighboring Gulf states or shift direction to other destinations through the corridor. During the so-called European migrant crisis, Bangladeshis became the top nationality represented among the incoming flows. In 2017, the first Bangladeshis (of approximately three thousand) arrived by boat in Italy from Libya, reporting after their arrival that they had been promised jobs by labor contractors who likely moved them through Tunisia to Libya and then across the Mediterranean on boats. Thus, even as the precarious status of contract labor in the Gulf shifts destinations of Bangladeshi emigrants, the existence of a mass migration corridor leaves in place networks that can be leveraged for large numbers of Bangladeshis leaving the country.

Insecurity and the Bangladeshi Climate Migrant

Because it is already prone to cycles of flooding along the Jamuna, Padma, and Meghna Rivers, Bangladesh is a key location of interest for international reporting on sea level rise and climate-driven migration, especially among U.S., European, and international reporters. Faced with rapid-onset flooding inundating homes and farmland, there are many accounts of individuals and communities facing direct displacement due to weather events. Journalists, policy makers, and climate scientists generally concur not only that Bangladesh is increasingly vulnerable to the effects of cyclones and flooding, but that sea level rise threatens its territory and population. The impacts of these changes are in progress, as hundreds of thousands face temporary or permanent displacement on an annual basis due to flooding. The Internal Displacement Monitoring Centre estimates that 946,000 people were displaced in Bangladesh in 2017 due to weather events. In 2007, the World Bank, citing Bangladesh government data, suggested that the capital Dhaka receives an estimated three hundred thousand to four hundred thousand migrants annually.[13] More recent estimates put the inflows to Dhaka at about six hundred thousand annually. By all estimates, Dhaka is rapidly becoming one of the world's megacities. According to sociologist Rita Afsar, internal migrants to Dhaka tend to be better educated, better off financially, and more likely to have significant kin obligations, while transnational migrants to the Gulf tend to have greater debts, have lower literacy rates, and be unmarried.[14]

When weather events precipitate environmentally influenced migration in Bangladesh, the options available for destinations remain substantially similar to those established in the 1980s. The concept of the "climate refugee" had not yet been invented when, in 1988, the first of three major flooding events of Bangladesh's postindependence era brought the country to a standstill. A massive cyclone produced floods that killed over six thousand people and wiped out 70 percent of the country's agricultural production. Large floods followed again in 1998 and 2004; these weather events resulted in an estimated 75 percent of the landmass being inundated with water. In the 1998 floods, thirty million Bangladeshis were displaced. Many people displaced by extreme flooding events may be able to return or to resettle locally, but when Bangladeshis seek to escape debt collectors or situations of persistent flooding, they often turn to Dhaka or to other large cities, like Chittagong. If climate change increases the frequency and number of Bangladeshi migrants, it is possible that the established domestic and international pathways will swell in numbers. Projections based on past Intergovernmental Panel on Climate Change (IPCC) estimates of Bay of Bengal sea level rise suggest significant future impacts. The more alarming predictions suggest that a three-foot rise in sea levels predicted by 2100 may swallow up to 20 percent of national territory, cause mass inland flooding and salinity-induced cropland losses, and displace large numbers of people (ranging from seventeen million to thirty-five million).

Given the dire predictions, Bangladesh has been an epicenter not only for reporting on purported climate refugees but also for speculation about mass migration, border conflict, ethnic violence, and Islamist terrorism, which, in the worst-case scenarios, appear explosive for all of South Asia. However, such speculation rarely involves close analysis of how environmental changes relate to the political economy of migration, which has been affected by an emphasis on export-oriented manufacturing located in the cities. In a widely cited report prepared for U.S. defense officials on climate change as a "threat multiplier," CNA's retired military consultants repeatedly mention Bangladesh as an epicenter of global climate migration.[15] Since 2008, the nonprofit CNA has commissioned a series of reports on water, famine, and other potential causes of security crises in the country. In 2014, it held a scenario-based exercise simulating a regional water crisis attended by A. N. M. Muniruzzaman as well as former government officials and water experts from India, Pakistan, the United States, and China. CNA's climate risk narratives suggest that Bangladesh is the most likely location of a future climate-driven U.S. military intervention. Categorizing Bangladesh as a high-risk and low-resiliency state, a recent report by CNA on climate-induced state instability identifies water system issues, famine, and weather events as reasons why Bangladesh may destabilize the entire region, making it a site of interest for U.S. military planners: "India's relationship with Bangladesh is one important consideration for the U.S. Since neighboring Bangladesh is fragile and important strategically, it may be a candidate for future U.S. assistance. The U.S. might be motivated to intervene there because that nation could harbor Islamic extremists and also because of its potential for conflict with India."[16]

Security analysts in both Europe and South Asia confirm similar threat assessments with their own apocalyptic speculations about climate-generated Bangladeshi displacement. Writing in *Foreign Policy* magazine, international relations professor Asma Khan Lone posits this argument:

> With survival at stake, most Bangladeshis will have to make their way into other territories—Assam in northeast India being a traditional destination. The influx into the region has led to a greater scramble for its limited resources sparking frequent violence. This has led to recurrent communal tensions, which could be exploited by

> trans-national elements such as ISIS or al-Qaeda with the latter citing widespread violence against Muslim Bengali migrants in 2014 as one of the causes (along with Kashmir) for establishing its local off-shoot, AQIS (al-Qaeda in the Indian Sub-continent). All these matters could come to a head, providing for a lethal ignition in the region.[17]

With India reinforcing its border with Bangladesh using armed troops and a new fence, and with the genocide against the Rohingya that has increased migration pressures on Bangladesh, several media and policy discussions of mass climate migration suggest the risk for regional border crises if climate change increases the numbers of displaced people. Although there is little specific discussion of how flooding would affect the transborder dynamics of communal conflict, a number of security observers have raised the specter that climate change will drive religious nationalism and Islamism. The German Advisory Council on Global Change in 2008 published a fictional scenario-based exercise in which policy makers are asked to respond to migration driven by sea level rise in Bangladesh as people flood into neighboring countries. The scenario script assumes that resource scarcity will drive competition organized along ethnic lines. Here is a summary of the scenario published by the Bangladesh Institute of International and Strategic Studies and the NGO Saferworld:

> Climate change will lead to a greater influx of migrants from Bangladesh into India, particularly the Indian states of Bihar, West Bengal and Assam, intensifying competition over resources between migrants and residents. This, in turn, could take the form of ethnic/religious conflict between Indian Hindus and Muslims from Bangladesh. . . . Local tensions could translate into diplomatic confrontation between the two countries if the Bangladeshi migrants are perceived as a security risk by the Indian government. At the end of this scenario, violent conflict is a real possibility. . . . The political conflict between the two states escalates; India threatens Bangladesh with "humanitarian intervention" on the pretext that the environmental migrants represent a terrorist threat.[18]

Religion operates as a shorthand for conflict in such speculative assessments, emerging as a vehicle of instability only when triggered by population-driven scarcity. In the process, Bangladesh is rendered as an emerging security threat because its designations as resource scarce, overpopulated, and poor intersect its potentials for religious violence and state collapse. Such thinking participates in a broader post-9/11 discourse that depicts Muslims as representatives of ingrained cultural differences that create a social tendency toward conflict and extremism.[19] The racial character of such discourses is clear not only in the general security mapping of risks focusing on Muslim-majority states but also in stereotypes of overpopulation and transborder invasion by downtrodden migrants. As such, the Bangladeshi climate migrant becomes an icon of the interlinked political, social, and ecological risks of the present. The policy-maker summary of the German advisory council report explains the regional and global risks of South Asian climate pressure in terms of a chain of events leading from migration to conflict, as it increases "the likelihood of conflict in transit and target regions." Loss of water and arable land, the report continues, increases the scale of environmental migration; although much of this will be South-South migration, "Europe and North America must also expect substantially increased migratory pressure from regions most at risk from climate change." Focusing on how "the annual monsoon will affect agriculture, and sea-level rise and cyclones will threaten human settlements around the populous Bay of Bengal," the summary concludes, "these dynamics will increase the social crisis potential in a region which is already characterized by cross-border conflicts (India/Pakistan), unstable governments (Bangladesh/Pakistan) and Islamism."[20]

The more paranoid accounts of climate-generated conflict in Bangladesh signal the potential for outmigration to ignite a regional war. Public health researcher M. Sophia Newman claims that the border situation is already on the brink of ethnic conflict echoing Rwanda. Regardless of what one thinks of the accuracy of this historical reference point, the argument appears to depend on assumptions about scarcity fueling social violence. Newman follows evolutionary biologist Jared Diamond to make Malthusian resource conflict the trigger event for the coming South Asian genocide:

> In rural areas, where environmental tensions can be most clearly felt, they arguably already are. For example, Bangladeshi Hindus were the targets of a two-month long spate of attacks in December 2013 to February 2014. The proximate causes were Islamic fundamentalists' anger over a contested election and war crimes tribunal; the violence was widely decried by moderate Muslims. But the violence is part of a long-running persecution of Hindus, and victims noted that some attackers appeared mostly interested in grabbing Hindu land and property. While political losses were the spark, poverty and resource

scarcity were the dry tinder for the attacks. The attacks looked much like the circumstances that preceded the Rwandan genocide of 1994—a conflict caused in part by resource scarcity, per scientist Jared Diamond's seminal book *Collapse*. Ongoing climate change threatens to increase the potential for Muslim attacks on Hindus inside Bangladesh. As Nazmul Hussain, a Bangladesh Army staff officer deployed with a peacekeeping force to Rwanda during the 1994 genocide, says, "I can see a similarity about potential threats in Bangladesh like Rwanda. . . . There is an ominous sign of potential outbursts anytime." If the threat is from Muslims to Hindus inside the country, though, the dynamic is reversed across the border. To speak of emigration to India is anathema to Bangladeshi tastes, despite ample evidence of long-standing Bengali Muslim migration into India's West Bengali, Tripura and Assam states, motivated by a quest for arable farmland. . . . India's newly elected prime minister, Narendra Modi, has publicly voiced his acceptance of migrants from Bangladesh—but only if they are Hindu. An outcry for inclusion from India's leftist politicians notwithstanding, the stipulation reflects long-standing Indian discrimination against Muslims. It may, in the end, cause international difficulties impacting millions.[21]

Vulnerability and Recognition

If the mapping of Muslim-majority state collapse configures the racial specter of instability at the global level of security narratives, reporting on the quotidian effects of climate change on rural Bangladeshis fills in the picture of vulnerability attributed to poverty and resource scarcity in the coastal areas of the Bay of Bengal. Empirical accounts of environmental displacement in Bangladesh differ substantially from abstract risk scenarios narrated by northern governments and security think tanks, all of which anticipate mass international crossings. The reporting of Gardiner Harris in the *New York Times*, for example, gives a sense of why much of the migration remains internal, as displaced individuals and groups attempt to reconstruct livelihoods under conditions of spatial alienation. Nonetheless, Harris's report, "Borrowed Time on Disappearing Land," recycles familiar colonial tropes of child slavery and poverty that echo much older colonial rescue narratives. Harris gives a detailed story of one family displaced in Khulna district, just north of the Sundarbans—the biodiverse mangrove delta on the Bay of Bengal:

> When a powerful storm destroyed her riverside home in 2009, Jahanara Khatun lost more than the modest roof over her head. In the aftermath, her husband died and she became so destitute that she sold her son and daughter into bonded servitude. . . . Ms. Khatun is trying to hold out at least for a while—one of millions living on borrowed time in this vast landscape of river islands, bamboo huts, heartbreaking choices and impossible hopes. . . . The poverty of people like Ms. Khatun makes them particularly vulnerable to storms. When [Cyclone] Aila hit, Ms. Khatun was home with her husband, parents and four children. A nearby berm collapsed, and their mud and bamboo hut washed away in minutes. Unable to save her belongings, Ms. Khatun put her youngest child on her back and, with her husband, fought through surging waters to a high road. Her parents were swept away. . . . The couple eventually shifted to the roof of a nearby hut. The family reunited on the road the next day after the children spent a harrowing night avoiding snakes that had sought higher ground, too. They drank rainwater until rescuers arrived a day or two later with bottled water, food and other supplies. The ordeal took a severe toll on Ms. Khatun's husband, whose health soon deteriorated. To pay for his treatment and the cost of rebuilding their hut, the family borrowed money from a loan shark. In return, Ms. Khatun and her three older children, then 10, 12 and 15, promised to work for seven months in a nearby brickmaking factory. She later sold her 11- and 13-year-old children to the owner of another brick factory, this one in Dhaka, for $450 to pay more debts. Her husband died four years after the storm.[22]

Although this account dramatically highlights the sale of children as bonded laborers, it details a story that will be familiar more generally among accounts of economic hardship in South Asia. Following the death or disability of a male laborer, family members struggle to obtain resources and capital sufficient to cover debt; in the end, they make the difficult decision to relocate children out of the natal home in hopes that their urban remittance can sustain the family unit. If such decisions represent what Harris calls the "heartbreaking choices and impossible hopes" of environmental migrants, they must also represent broader problems of economic inequality, industrial and export-led development, and loan predation in South Asia (a problem that at times has been exacerbated by NGO resilience schemes, such as microcredit). Nonetheless, the story gives a useful spatial account of how floodplain-to-urban migration patterns enlarged by climate change are likely to stick to established routes. Bangladesh's export-oriented development model make Chittagong and Dhaka more likely destinations of cash-seeking environmental migrants than India's Assam border crossing, which remains a preoccupation of many security experts.

While news reporting on climate migration offers a mixed portrait of local, regional, and global processes that may influence migration, U.S. pundits writing in the same publications at times engage in gendered, colonial tropes of the Asian city, invoking Malthusian discourse on population as the main social pressure creating personal tragedies. Writing in the *New York Times*, Nicholas Kristof's opinion column mentions the role of climate change in the Syria and Darfur wars and cites Christian aid groups to claim that climate change increases forced marriage of adolescent girls in Bangladesh. Kristof's description of suffocating population concentration in Dhaka frames the tropicalizing narrative of urban congestion:

> To stroll here in the mists of early morning is to navigate an obstacle course of makeshift beds and sleeping children. Later the city's steamy roads and alleyways clog with the chaos of some 15 million people, most of them stuck in traffic. Amid this clatter and hubbub moves a small army of Bengali beggars, vegetable sellers, popcorn vendors, rickshaw drivers, and trinket salesmen, all surging through the city like particles in a flash flood. The countryside beyond is a vast watery floodplain with intermittent stretches of land that are lush, green, flat as a parking lot—and wall-to-wall with human beings. In places you might expect to find solitude, there is none. There are no lonesome highways in Bangladesh.[23]

Reversing colonial depictions of nature as a place of refuge from industrial modernity,[24] the essay frames population concentration as the inevitable lot of the Bangladeshi poor who are vulnerable to climate change. Kristof figures the Bangladeshi internal migrant as a kind of natural force, moving across urban and rural space "like particles in a flash flood." Tracking human mobility from Kristof's tourist perspective—which reverses the rural-to-urban pathway of the migrants themselves—the essay turns from the romantic depiction of the urban intimacy of working-class entrepreneurs to the "wall" of human beings on the move in the rural, flood-affected areas.

This emphasis on unending population, recapitulating depictions of colonized hordes overwhelming colonial observers in their travel narratives, demonstrates no attention to the localized processes of decision making about migration. Even when weather events trigger sudden migration, such migration decisions are usually based on a complex calculus of social and economic factors and are clustered along migration routes and networks that are both internal and external to Bangladesh. As such, no migration is purely climate migration, and existing economic inequalities created by the carbon-fueled regimes of trade globalization have much to do with the high volume of migration in climate-affected regions. Kristof's narrative nonetheless demonstrates how the histories of such regions are easy to write over with deterministic environmental narratives, suggesting that Earth system processes overwhelm the agency of those affected by their outcomes.

It is useful in the face of neocolonial narratives of environmental determinism to take note of another set of transformations affecting Khulna's outmigration: moves by development agencies to promote climate adaptation schemes that in fact accelerate displacement. In her research on displacement in Khulna district, Kasia Paprocki details how, in the midst of land grabs by large landholders that displace coastal agrarian populations, development agencies have promoted shrimp aquaculture as an alternative to the coastal farming practices preferred by small agriculturalists. Describing a process of "anticipatory ruination," Paprocki demonstrates how the vision of Khulna as a dying place affected by rising seas allows landholders and development agencies to promote aquaculture as an alternative to traditional rice cultivation, in the process detaching local residents from existing livelihoods and favoring large concentrated schemes for development.[25] These schemes are based on an idea of the uninhabitability of the coastal region and a vision of development that requires coastal agrarian populations to develop new skills and to migrate for urban industrial work elsewhere. And yet they actually *promote* the degradation of the land base as aquaculture operations pump in salt water to the coastal plain, killing trees and other vegetation in the process. In a gothic reversal of John Locke's vision of settler colonial plantation development, shrimp aquaculture in Khulna seeks to accumulate wealth based on accelerating the transformation of the landscape into wasteland in order to both hasten human outmigration and clear the land for large property owners to farm shellfish.

The Future of Adaptation: Climate Refugees as a Security Resource

Although some international journalism recycles long-standing racial depictions of South Asians, stereotypes about climate migration remain durable in part because they help to underscore specific diplomatic and policy appeals of the state. The Bangladeshi government is aware of the emerging discourse on climate migration and appears to find it useful in its policy strategies around migration and security. Like small island states, the Bangladeshi government has taken steps to reduce the impacts of climate change, including establishing an early warning system for flooding. Although significant inequalities keep over 40 percent of the population in poverty and many more economically vulnerable, state officials tend to stress resiliency, arguing that industrial development, climate adaptation schemes, and the expansion of health care have significantly reduced the mortality impacts of flooding over the past decade.

However, even though they have often attempted to counter stereotyped images of Bangladeshi poverty and instability, Bangladeshi government officials also overemphasize the potential that climate change will produce mass outmigration. This reflects a set of state agendas that favor the interests of middle-class Bangladeshis, particularly as they are viewed as a potentially greater source of foreign dollar remittances than are poor migrants. Two pressures appear to influence the strategic use of climate security discourse by Bangladeshi climate diplomats. First, GDP growth combined with a change in the definition of national wealth at the World Bank—driven in part by attempts to internalize natural resources into wealth estimates—has taken Bangladesh out of the group of extremely poor countries designated by the International Monetary Fund, placing more pressure on the Bangladeshi government to advocate for aid from international NGOs and emphasize its vulnerability to climate change. This change in status occurred despite the fact that Bangladesh continues to have the lowest income levels of any country in South Asia, meaning that some receiving countries in the Gulf boast average incomes of seventy times the average Bangladeshi income.[26] Second, outmigration to the West is most likely to benefit educated middle-class Bangladeshis who seek professional employment abroad. The state has not prioritized aid to silt island dwellers and other informal settlements in flood risk areas, hoping to discourage settlement in areas subject to cyclical floods, yet it seeks expanded international resettlement to the rich countries in its negotiations of international migration agreements, such as the Global Compact for Migration. Conditions of unequal development and aid may help explain why Bangladeshi officials periodically raise dire warnings about outmigration, which may in fact benefit educated middle-class Bangladeshis who are not specifically environmental migrants rather than Bangladeshis who are fleeing flooding within the country or those who, affected by loss of land or crops, are seeking new employment in the Gulf.[27]

Sometimes, without specific rationale, government estimates trend significantly above scientists' estimates. In 2014, Tariq Kalim, Bangladesh's ambassador to India, estimated in the *New York Times* that as many as fifty million Bangladeshis could flee to neighboring countries by 2050—a figure (far beyond the usual claims of seventeen million displaced) that seems dubious given existing trends as well as the high proportion of migrants who are likely to remain internally displaced.[28] Such statements may work against state efforts to advertise an image of growth and modernity for Bangladesh. But they are a key rhetorical strategy in the government's push for resettlement commitments from the Global North. Bangladesh's Climate Change Strategy and Action Plan of 2009 demonstrates how the state at times portrays such migrants as potentially useful laborers: "It has been estimated that there is the impending threat of displacement of more than 20 million people in the event of sea-level change and resulting increase in salinity coupled with impact of increases in cyclones and storm surges, in the near future. The settlement of these environmental refugees will pose a serious problem for the densely populated Bangladesh and migration must be considered as a valid option for the country. Preparations in the meantime will be made to convert this population into trained and useful citizens for any country."[29]

Echoing development discourses that view agrarian peasants as in need of education and job training for the industrial economy located away from their present sites of habitation, such discourse on human capital development reflects the anticipatory clearing of the agrarian poor from the landscape that comes with climate adaptation planning. There is, however, no consensus in the literatures in geography or security studies that environmental scarcity produces conflict.[30] Consider the rise of autonomous mutual aid societies in the Carolinas and Puerto Rico that have filled the gap in the wake of failures by state agencies to redress hurricane impacts. Such efforts align with arguments that environmental disasters may produce cooperation rather than conflict, a line of thinking that the Left has long used to critique social Darwinism and other environmental or biological determinist theories of social action.[31] Taking account of this competing line of research, the IPCC's fifth assessment report claims that "evidence on the effect of climate change and variability on violence is contested." It even cites a growing body of scholarship detailing how climate adaptation and mitigation strategies are more likely to produce conflict. Furthermore, IPCC authors do not portray climate-influenced migration as a uniformly negative phenomenon. As discussed in chapter 1, the fifth assessment report's chapter on human security makes two interrelated points: (1) "migration and mobility are adaptation strategies," and (2) "indigenous, local, and traditional forms of knowledge are a major resource for adapting to climate change (robust evidence, high agreement)."[32] These statements view climate migrants not as bearers of instability but as embodiments of and resources for adaptation. As such, emphasizing cooperation over conflict interfaces with emerging climate security discourses in a different way, seeking to accumulate the knowledge and labor of climate-affected Indigenous groups and climate migrants for state- and interstate projects of adaptation.

Such claims help to develop a related discourse on climate migration that moves from Malthusian speculations of instability to romantic invocations of human resilience and capacity for overcoming adversity. Although this breaks with some of the more pernicious, militarized depictions of climate insecurity, such depictions of climate adaptation dovetail with other equally distorting visions of individual transcendence of adversity. These may in turn buttress adaptation strategies that envision the removal of agrarian populations from coastal areas as new capitalist enterprises are put in place. Such moves from the fear of population bombs to romantic invocations of resilient adaptation include photography and journalism focusing on the char dwellers of Bangladesh's Jamuna River. These river migrants have become famous for their ability to rapidly adapt farming, built structures, and social practices to a constantly shifting environment. The chars are small silt islands that flood and re-form with the ebbs and flows of water levels. Romantically depicting the heroic efforts of adaptation among the char dwellers, *National Geographic* describes life on the chars as such:

> These islands, many covering less than a square mile, appear and vanish constantly, rising and falling with the tide, the season, the phase of the moon, the rainfall, and the flow of rivers upstream. Char dwellers will set out by boat to visit friends on another char, only to find that it's completely disappeared. Later they will hear through the grapevine that their friends moved to a new char that had popped up a few miles downstream, built their house in a day, and planted a garden by nightfall. Making a life on the chars—growing crops, building a home, raising a family—is like winning an Olympic medal in adaptation. Char dwellers may be the most resilient people on Earth.[33]

The 2011 *National Geographic* article on the char dwellers, "The Coming Storm," invokes the types of tragedy explored in Harris's *New York Times* reporting but quickly turns to arguing that Bangladeshis' protracted vulnerability leads to an unusual ability to adapt. It does so by configuring security officials' speculations on the climate-driven disasters facing the country as part of a longer history of protracted tragedy afflicting the nation. Don Belt's text for the article references the natural, constitutional resilience of the bodies and social formations of coastal migrants:

> Such a catastrophe, even imaginary, fits right in with Bangladesh's crisis-driven story line, which, since the country's independence in 1971, has included war, famine, disease, killer cyclones, massive floods, military coups, political assassinations, and pitiable rates of poverty and deprivation—a list of woes that inspired some to label it an international basket case. Yet if despair is in order, plenty of people in Bangladesh didn't read the script. In fact, many here are pitching another ending altogether, one in which the hardships of their past give rise to a powerful hope. . . . The one commodity that Bangladesh has in profusion . . . [is] human resilience. Before this century is over, the world, rather than pitying Bangladesh, may wind up learning from her example.

Jonas Bendiksen's photos accompanying the *National Geographic* essay serve to illustrate the purported ability of the char dwellers to flexibly adapt to rising waters. This isn't a generic claim, as the images literally depict people living in inundated houses. Rather than maintaining the tragic mode for representing flooding that must cause sustained difficulty in the household labor needed to reproduce the family unit, the image slideshow depicts the char dwellers as modeling efforts needed for climate adaptation—in this case displaced from any responsibility of polluters or the state as individuals develop makeshift water avoidance strategies. Several of the photos employ the convention of visually dividing the scene into two areas via an infrastructural cut through the landscape, such as a road or a waterway, depicting the movement of migrants through the landscape or creating a temporal dimension of transition through the contrast between the two sides of the frame. Such cues are aimed at a viewer presumed to be outside Bangladesh, whose vision of the images is framed by captions containing brief ethnographic descriptions of the people and the country, accompanied by algorithmically targeted hypertext ads for consumer goods and a lower banner connecting to other stories, such as an article concerning dinosaur fossils.

Hot, flat, and crowded streets of Dhaka, in Don Belt, "The Coming Storm," *National Geographic*, May 2011, www.nationalgeographic.com/magazine/2011/05/bangladesh. Photo by Jonas Bendiksen.

Some images in the slideshow include views of overpopulation reminiscent of Kristof's depictions of urban concentration and rural mass migration; one photo of riverboats at Dhaka's main terminal is accompanied by the caption "Overflowing with People," describing Dhaka as "one of the world's most populated cities." Another photo's caption contends that "Dhaka swells with migrants from the flood- and storm-ravaged countryside," as it visually contrasts the left side of a street, loosely populated by rickshaws, pedestrians, and shops, with the densely packed right side, full of barefoot men standing atop prayer mats, the street (and photo) divided by a median. Evoking the connection between Malthusian discourses of overpopulation and the view of Islam as a node of potential climate security risk, the caption claims, "The streets of Dhaka absorb the overflow crowd from a mosque celebrating the end of Ramadan."

But more characteristically, the images retreat from sensationalizing population bombs and invoke self-uplift as the aftermath of disaster. Without always identifying the locations or individuals depicted, these images depict char dwellers and other Bangladeshis affected by rising waters—especially via images of the persistence of women and children—as evidence of the possibility of social reproduction after floods. The slideshow opens with a rain-soaked image of a child, titled "Resilient Spirits" (see cover). Clothed only in shorts, the youth walks barefoot down a drenched

Resilient Spirits, in Don Belt, "The Coming Storm," *National Geographic,* May 2011, www.nationalgeographic.com/magazine/2011/05/bangladesh. Photo by Jonas Bendiksen.

coastal road in the midst of a torrential downpour, trees blowing in the background. The image, taken during a heavy monsoon, configures the child as well as an adult villager in the distance as specters whose blurred moving outlines amidst the mud, rain, and fog suggest both the possible vulnerability to weather and the potential for transcendence. This image, arranged with the road bisecting the landscape, is echoed by another romantic image of displacement that similarly centers transit infrastructure in a changing landscape. Titled "City Bound," this next photo depicts a number of men sleeping atop a moving train that transits lush green farmland toward the interior. The caption clarifies that these passengers are returning to Dhaka after visiting home villages, crossing rice paddies that are currently thriving but that are close to others threatened by salinization as the seas rise into the river delta. Concluding this journey from floodplain to city, the photo essay goes on to depict the crowded slums of Dhaka as "bursting with environmental refugees."

From here, Bendiksen's images move from the tragedy of migration to the triumph of resilience, which is evoked in depictions of the changing spaces of work, home, and worship under flood conditions. In an image titled "Keeping a Country Afloat," waterlogged rice paddies are revamped as salt ponds for shrimp and crab harvesting. Without describing the struggles between large enterprises and small farmers in the conversion to aquaculture

City Bound, in Don Belt, "The Coming Storm," *National Geographic,* May 2011, www.nationalgeographic.com/magazine/2011/05/bangladesh. Photo by Jonas Bendiksen.

At a Breaking Point, in Don Belt, "The Coming Storm," *National Geographic,* May 2011, www.nationalgeographic.com/magazine/2011/05/bangladesh. Photo by Jonas Bendiksen.

Keeping a Country Afloat, in Don Belt, "The Coming Storm," *National Geographic*, May 2011, www.nationalgeographic.com/magazine/2011/05/bangladesh. Photo by Jonas Bendiksen.

described in Paprocki's research, the image evacuates the agrarian poor from the scene of adaptation in order to emphasize the entrepreneurial success in sustaining food production amid rising waters. This image repeats the visual composition of a landscape divided by transit infrastructure—again a road—that swirls from front to back through the flooded farms. This reference to aquaculture suggests the influence of national and international nonprofit organizations promoting climate adaptation in the article's construction of national resilience. Two images (not displayed here) focus on projects sponsored by two nongovernmental organizations, Practical Action and Friendship, depicting respectively a floating garden and a hospital ship financed by the Emirates Airline Foundation. This last example shows some of the connections of the oil-fueled Gulf labor economy that flow back into adaptation schemes, especially as several Gulf states have aid investments in Muslim-majority countries, including Bangladesh, through initiatives such as the Islamic Development Bank.

Additional images show how adaptation strategies frame the experience of domesticity. One image, titled "Home for the Moment," focuses on what is identified as the Uddin household in their home, where one foot of water is unable to prevent a woman from cooking the evening meal upon the family's stone hearth. As the men of the household carry two babies above the waterlogged floor, several women are seated around the hearth, observing the cooking amidst large sacks of grain and an assortment of cooking pots. Explaining that the family had recently moved to escape flooding, the caption claims that they intend to dismantle and rebuild once again. In this invocation of reproductive labors involving the family's efforts at provision and childrearing, the climate refugee's resilience signals futurity in the face of threats to settlement. Whereas older development narratives tended to emphasize the potentially destructive aspects of children as avatars of overpopulation, this image suggests a hopeful survivalism and the potential for adaptation to aqueous habitation. Portraying two children who have climbed several feet up posts to stay elevated in a flood event, another image titled "High and Dry" suggests that adaptation can be reproduced as children integrate it into everyday life practices. The caption clarifies that "char dwellers . . . are used to such calamities, which are on the increase." Another two images depict the process of mobile domiciling, showing a structure capable of disassembly and movement as the river path reorients the amount of available land among the chars. In this case, the images show a mosque being carried by a group of people crossing the char and then reconstructed, with one caption claiming that the afternoon move was fast enough to have the mosque operating by the time of evening prayers. These images directly invoke the Muslim-majority demographics of Bangladesh without resorting to security discourses about the potential for extremism; here, the efforts to maintain a mobile mosque are integrated into a broader set of representations that invoke the romantic potential for adaptation and the national capacity for resilience in the face of disaster.

Yet in Belt's narrative for *National Geographic,* such small-scale portraits of human resilience blur into more established narratives of large-scale crisis. Toggling between a Malthusian nightmare and liberal adaptation fantasy, the article suggests that the hope offered by Bangladeshis' constitutional capacity for adaptation is circumscribed by the potential for mass destruction. In his quotes for the article, Major General Muniruzzaman (Ret.) dubs climate migration "the largest mass migration in human history." Despite the article's emphasis on resilience, Muniruzzaman stresses two possible outcomes of mass migration due to climate change: (1) the sudden movement of people causes catastrophic security crises, or (2) rich countries establish migration policies that could act as a safety valve for such population pressures. In the second of these scenarios, Bangladeshi state officials and development experts appear to suggest the potential of mining the human capital of climate refugees:

Home for the Moment, in Don Belt, "The Coming Storm," *National Geographic,* May 2011, www.nationalgeographic.com/magazine/2011/05/bangladesh. Photo by Jonas Bendiksen.

High and Dry, in Don Belt, "The Coming Storm," *National Geographic,* May 2011,www.nationalgeographic.com/magazine/2011/05/bangladesh. Photo by Jonas Bendiksen.

Nothing lasts on Sirajbag, in Don Belt, "The Coming Storm," *National Geographic*, May 2011, www.nationalgeographic.com/magazine/2011/05/bangladesh. Photo by Jonas Bendiksen.

Seeking Higher Ground, in Don Belt, "The Coming Storm," *National Geographic*, May 2011, www.nationalgeographic.com/magazine/2011/05/bangladesh. Photo by Jonas Bendiksen.

> If ten million climate refugees were ever to storm across the border into India, Maj. Gen. Muniruzzaman says, "those trigger-happy Indian border guards would soon run out of bullets." He argues that developed countries — not just India — should be liberalizing immigration policies to head off such a chilling prospect. All around Bangladesh[,] bright, ambitious, well-educated young people are plotting their exit strategies. And that's not such a bad idea, says Mohammed Mabud, a professor of public health at Dhaka's North South University and president of the Organization for Population and Poverty Alleviation. Mabud believes that investing in educating Bangladeshis would not only help train professionals to work within the country but also make them desirable as immigrants to other countries — sort of a planned brain drain. Emigration could relieve some of the pressure that's sure to slam down in the decades ahead. It's also a way to bolster the country's economy; remittances sent back by emigrants account for 11 percent of the country's GDP. "If people can go abroad for employment, trade, or education and stay there for several years, many of them will stay," he says. By the time climate change hits hardest, the population of Bangladesh could be reduced by 8 to 20 million people — if the government makes out-migration a more urgent priority.

If the essay, following Muniruzzaman's depiction of India as the aggressor country, avoids stereotyped security depictions of Islamic militancy, it nonetheless conjures the potential for scarcity-driven mass migration and conflict as the only imaginable outcomes of climate-driven flooding in Bangladesh. The article thus dovetails with the development plans of the state and NGOs to support outmigration. The imperative of the Bangladeshi government to transform peasant agriculturalists into a global laboring force in order to bolster a claim to environmental asylum presents a fascinating glimpse of how Malthusian views of conflict and sovereignty might be carried to their logical conclusions by climate-affected states. This vision of the planned depopulation of Bangladesh suggests that the true adaptation that must occur is to the regime of transnational capital and trade. The Bangladeshi peasant laborer is vulnerable to climate change if that laborer is localized — that is, dependent on the land. Climate displacement matters less if the peasant can abstract labor from land.

This last point is key, both for the fates of those affected by environmental change and for their relation to the state. Part of the agrarian vulnerability reproduced in places like the Bengal delta emerges from the intentional forgetting that occurs with colonial development of riverine environments. In important work on the ecological history of Bengal, Debjani Bhattacharya argues that the act of forgetting the river's shaping of the city was central to the building of Calcutta as a colonial capital over a shifting deltaic environment. Hydrological control was central to filling in areas in which the river changed direction, allowing for the eastern side of the city to be built up and to drive its role in British commercial enterprise.[34] Naveeda Khan, who is doing important ethnographic research on char dwellers along the Jamuna, notes that the community of itinerant farmers she studies dates to the 1930s, when the river shifted westward and displaced hundreds of thousands of villagers. This history of environmental change adds localized resolution to the specifics of human-environment interaction, demonstrating the long-standing imbrication of human bodies in the riverine environment in eras prior to what appears today to be the acceleration of flooding events. In the intervening decades since the 1930s, state efforts to build embankments have been ineffective or counterproductive, resulting in devastating floods such as those that took place in 1988. Following this, char dwellers were left out of government flood planning because shifting char lands were not seen as worthy of state protection. In the absence of state action to enforce flood protection or property laws on char lands—and in defiance of engineering decisions that mitigate flood risk to the mainland at the expense of the char islands—char-dwelling groups have developed unique cultural forms and have been able to maintain village identity autonomously. Exceptional representations of char dwellers as romantic figures of climate resilience foreclose the agency of the river and the complex set of embodied interactions through which societies have emerged on shifting silt islands.

Attempting methodologically to center not just the inhabitants of char islands but also the phenomenal world of the river itself, Khan notes the difficulty in discussing what aspects of riverine life are the result of climate change and what are longer-term aspects of the lifeworld:

> While Bangladesh is undoubtedly affected by climate change . . . it is difficult to know for sure what comprises climate change within the river system. The usual fear related to rivers is that of flooding. With global warming, the increased precipitation from temperature rise is projected to produce more rainfall causing the Jamuna and other rivers to overflow more often and more unexpectedly. At present the

> standing rainfall data does not point definitively towards a rise in temperature within the country or even in the region and therefore to increased floodwaters. Yet there is still widely expressed concern about floods arising from global warming. I would argue that this concern is in part about the growing inability to anticipate floods . . . [whose] cyclicality does not seem to hold at present. Furthermore, this frequently articulated concern about floods arising from global warming is expressive of an anxiety among climate scientists in Bangladesh that regional processes and local variability underlying rainfall have not been sufficiently integrated . . . to be able to say anything definitively about climate change.[35]

By disentangling the lifeworld of the river and the char dwellers from the perspective of the Bangladeshi state, Khan distinguishes some of the elite prerogatives of climate security from the phenomenal complexities of inhabiting changing interspecies landscapes and ecologies. This involves setting aside an aesthetic of nature that is overdetermined by securitizing narratives of apocalyptic disaster. For Khan, "the enframement of one's environment by the concept of disaster makes difficult the ability to see nature as anything other than pregnant with risks. Consequently, this powerful disaster-laden perspective on nature has ramifications for disinterring climate from weather."[36] Integrating discussion of the complexities and uncertainties of climate-affected water systems, shifting human habitation, and the complex movements of humans and other species requires us to attend both to the ways that oil-fueled capitalism has remade the international order and to the emerging ways that ecological processes map on to the existing geopolitical relations established by it.

Reframing the Mobilities of Climate

The figure of the climate refugee in Bangladesh thus reveals some significant changes in the politics and aesthetics of climate change in the aftermath of the collapse of international efforts to mitigate carbon emissions. First, it suggests that discourses about population, migration, and conflict signal the emerging shift from international to national efforts to remedy climate change, and from mitigation to adaptation and security frameworks. Second, it reflects how visions of both displacement and adaptation are emerging in ways that shift responsibility for displacement from capitalists and states to individuals who are most affected by weather disasters.

Finally, it suggests that the hopeful focus on adaptation to climate change—a kind of overcoming of disability—configures migrant knowledge as a potential solution to crisis, even as states and analysts continue to invoke the Malthusian crisis scenarios of the mass migration frame.

This ambivalent move between large- and small-scale narratives of climate migration—from the mass migration scenarios of state conflict and collapse to the intimate portraits of individual effort and transcendence—reflects racialized dynamics in the politics of knowledge concerning climate change and conflict. This situation holds significant lessons for emerging discourses about time, space, scale, and agency in research on the Anthropocene and in environmental humanities research. Indeed, Bengal as a region encompassing Bangladesh and the Indian state of West Bengal has become an important location for recent theorizing of the Anthropocene in part because of the role diasporic scholars from the region have played in describing the relevance of climate change's geophysical processes to politics, art, and history. Although climate migration discourse has not been at the center of these academic conversations, it is necessary to think about the interrelation between the idea of the Anthropocene and the icon of the climate migrant in order to understand how race, empire, and knowledge are being constructed.

The article that launched discussions of the Anthropocene among humanists in the United States, Dipesh Chakrabarty's "The Climate of History: Four Theses," was originally published in the journal *Baromas* in Kolkata before its translation for the English-language journal *Critical Inquiry*. Reflecting on this article and the experience of writing about climate change from India and more broadly from the Global South, Chakrabarty describes an international imbalance in access to information and debate about climate change:

> Globalization—including questions about multinationals, money markets, derivatives and complex financial instruments, the net, the social media, and, of course, the global media—was a genuinely global topic that was discussed everywhere but global warming was not. And it also became clear who set the terms of the discourse. It was the scientists of nations that played a historical role in precipitating the problem of global warming through their emission of polluting greenhouse gases—for example, the United States, the United Kingdom, Australia and other developed countries—who played two critical roles: as scientists, they discovered and defined the phenom-

> enon of anthropogenic climate change, and as public intellectuals they took care to disseminate their knowledge so that the matter could be debated in public life in an informed manner. . . . Global warming is a planetary phenomenon. But as a subject of discussion, it seemed to be distributed very unequally in the world. The situation has changed somewhat in the last ten years—thanks in part to the increasing frequency and fury of extreme weather events in different areas of the world—but not substantially. What are the implications of this disparity in the distribution of information? It surely skews the "global" debate on climate change in more than one way. When governments come to global forums to discuss and negotiate global agreements on climate change, they do not come equally resourced with informed public discussions in their respective nations, while some governments, admittedly, do not even desire informed publics. More importantly, it means that our debates remain anchored primarily in the experiences, values, and desires of developed nations, that is, in the West (bracketing Japan for the moment), even when we think we are arguing against what we construe to be the selfish interests of "the West."[37]

Chakrabarty's critique certainly applies to the discourse about climate migration, which appears to center on the interests of the poorest climate-affected peoples in the Global South but actually recapitulates a number of colonial stereotypes about the deterministic effects of resource scarcity, in the process reinforcing anti-Muslim security discourses that stress the potential for climate migrants to produce climate wars. Anchored by the writings of security experts in northern states and in security NGOs, such arguments rehash long-standing mappings of world risk focusing on Africa and Asia as sources of global instability. At the same time, the role of the state in articulating such discourses might give us pause as to whether the deep inequalities of wealth that drive insecurity map so easily on the inequalities between states.

Although Chakrabarty's prior work had stressed the significance of considering postcolonial difference within accounts of the global,[38] "The Climate of History" argues that it is necessary to retain the human as a universal category despite such critiques of capitalism's notion of the human.[39] In defending the Anthropocene theorists' conception of the human as an environmental agent affecting planetary systems, Chakrabarty argues that it is necessary to focus on the human as a biological species and to consider how the deep histories of human social organization—including the transition to sedentary agriculture—unknowingly create massive changes in the geophysical form of the planet. This point about the unintentional violence of human environmental subsistence and settlement is the crux of Chakrabarty's claim that it is necessary to tell a new history of the human, not as an identity but as a form of universality that "escapes our capacity to experience the world."[40] Chakrabarty here conceives of environmental harms as externalities of human activity and social organization that can only be retrospectively accounted for through empirical scientific observation. This echoes the arguments of some environmental justice advocates and climate scientists discussed in chapter 1, who engage in the retrospective accounting of environmental harm. Chakrabarty ends the essay by invoking the figure of a human species that unintentionally provokes a planetary crisis, allowing it to sense its own collective agency:

> The anxiety global warming gives rise to is reminiscent of the days when many feared a global nuclear war. But there is a very important difference. A nuclear war would have been a conscious decision on the part of the powers that be. Climate change is an unintended consequence of human actions and shows, only through scientific analysis, the effects of our actions as a species. Species may indeed be the name of a placeholder for an emergent, new universal history of humans that flashes up in the moment of the danger that is climate change. But we can never understand this universal. It is not a Hegelian universal arising dialectically out of the movement of history, or a universal of capital brought forth by the present crisis. . . . Yet climate change poses for us a question of a human collectivity, an us, pointing to a figure of the universal that escapes our capacity to experience the world. It is more like a universal that arises from a shared sense of a catastrophe.[41]

Setting aside for the moment the equal possibilities that nuclear war could have been unintentional or that the extractive enterprises of the great acceleration were intentionally organized forms of accumulation, the role of the state in managing the development potential of the displaced climate migrant suggests that even if the figure of human universality may be useful for attending to large-scale planetary changes, it may have important distorting effects at the level of political economy, where generalizations about the effects of climatic processes mask how systemic inequalities create conditions in which the boundaries of livable and unlivable lives are drawn differently based on geographic, class, and national differences. The possibility that Bangladesh will use climate discourse in order to mobilize middle-class outmigration to the rich countries of the North must be understood within a longer context of its neoliberal approach to developing human capital as one of its export-oriented development strategies. So if, according to Chakrabarty's argument, a universal history of the human as an environmentally destructive agent is necessary for understanding the large-scale processes of climate change, it is also pivotal to disaggregate the human subjected to the precarious futures of global warming in order to understand some of the smaller local- and regional-scale networks that are established through the carbon economy's overlapping extractive and waste-generating processes. Put differently, planet-scale narratives of human agency are important in understanding certain processes of environmental change but may risk distorting how particular groups of people affected by floods and other disasters are likely to navigate the limited migration options generated by capitalist labor networks.

When attending to localized stories of environmental mobility caused by flooding, it becomes apparent that the deep time of environmental change is more complex and circuitous than the species-wide narratives of the human as geological agent. Amitav Ghosh, one of the most prominent climate change essayists today, foregrounds the agency and perspectives of subjects affected by environmental processes. Ghosh, the Bengali writer known for numerous novels, including *The Hungry Tide*, himself identifies as coming from a family of ecological refugees. The opening pages of his book *The Great Derangement: Climate Change and the Unthinkable* tell the story of his kin's displacement:

> My ancestors were ecological refugees long before the term was invented. They were from what is now Bangladesh, and their village was on the shore of the Padma river, one of the mightiest waterways in the land. The story, as my father told it, was this: one day in the mid-1850s the great river suddenly changed course, drowning the village; only a few of the inhabitants had managed to escape to higher ground. . . . It was this catastrophe that unmoored our forebears; in its wake they began to move westward, when they settled again on the banks of a river, the Ganges, in Bihar.[42]

Ghosh uses this accounting to describe a process of environmental recognition. Ghosh argues that environmental changes like the flooding of his family's village are the sites at which people develop a new type of recognition of the nonhuman world, an awareness of the inseparability of the inside and outside of the body as well as the dependence on the environment as a background set of conditions enabling (or potentially destroying) a life-world. Drawing on this description of the experience of environmental harm, Ghosh argues that the process of recognition involves accessing a previously unacknowledged past. Echoing Anthropocene discourses focusing on the retrospective observation of species-scale agency, Ghosh rhetorically scales this down to the site of the subject and describes crisis as an opening to environmental awareness: "These too are moments of recognition, in which it dawns on us that the energy that surrounds us, flowing under our feet and through wires in our walls, animating our vehicles and illuminating our rooms, is an all-encompassing presence that may have its own purposes about which we know nothing. It was in this way that I too became aware of the urgent proximity of nonhuman presences, through instances of recognition that were forced on me by my surroundings."[43]

Drawing on Ghosh's invocation of intimacies with the landscape, the lesson for understanding migration in an era of climate change is not so much that a new universal vision of human history is necessary but that the types of mobility networks available to people who experience weather disasters have everything to do with the outcomes of how environmental violence grafts on to the existing violences of racism, colonialism, and capitalism. Many climate-affected people have already been living in worlds charged by struggles both to inhabit the land and to continue forms of subsistence disfavored not just by the changing landscape but by the interests of capital and institutions. In such contexts, recognition of nonhuman agencies in the surrounding environment may help people navigate uncertain futures, even though when imposed by the state and development agencies they can compound the variety of external influences that already participate in the destruction of livelihoods and lifeworlds. From here, what is needed is not just recognition of the sublime shifts in scale, space, and time created by environmental processes but also attention to the differential needs and capacities of people affected by both extractive enterprise and environmental change.

CHAPTER FOUR

Weather as War

Race, Disability, and Environmental Determinism in the Syrian Climate War Thesis

One of the most widely publicized climate change studies of recent years was a paper published in the *Proceedings of the National Academy of Sciences* (*PNAS*) in 2015.[1] Colin Kelley, a climate scientist who was then a postdoctoral fellow at the University of California–Santa Barbara, cowrote the article with colleagues in the fields of international relations and earth sciences at Columbia University. Based on Kelley's data surveying rainfall patterns in the Fertile Crescent region of North Africa and the eastern Mediterranean, the authors argue that the 2006–9 drought in the region contributed to the outbreak of the Syrian war of 2011. The authors frame drought as a trigger event for rural-to-urban migration, which they argue brought about the conditions for protest and conflict. Furthermore, the authors assert that the events of the beginning of the war were made more likely due to decades of Syrian agricultural policy that overused subsurface water resources and created unsustainable farm dependency on water importation.

Since the publication of the article, titled "Climate Change in the Fertile Crescent and Implications of the Recent Syrian Drought," journalists, environmentalists, and security analysts worldwide picked up the story and began framing the Syrian uprisings, Syrian migration to Europe, and the Syrian war as evidence that Syria was experiencing one of the world's first "climate wars." The *National Geographic* article announcing the study reported that "a severe drought, worsened by a warming climate, drove Syrian farmers to abandon their crops and flock to cities, helping trigger a civil war that has killed hundreds of thousands of people."[2] Although there has been a mixed reception of the article among both climatologists and international relations scholars,[3] reporting on the potential for climate change to cause transnational conflict accompanied by mass migration has only increased in the intervening years, with Syria remaining a key site representing the kind of social breakdown possible due to climate change. The story was picked up by major news outlets, including the BBC, NPR, and the *Washington Post*. It quickly became a topic of editorializing by military officials and security experts, who began to sound the alarm that climate change threatened globally distributed conflict in the twenty-first century. Although some environmental NGOs had suggested as early as the 1980s that growing water scarcity was the cause of resource conflicts,[4] the *PNAS* article put climate-induced drought front and center in public debates over the effects of climate change, reaching an international audience of experts and laypeople.

The *PNAS* article inaugurated public circulation of the Syrian climate war thesis, a narrative that presents drought as a trigger of rural-to-urban migration that unleashed social and state breakdowns by creating an opening for Islamist groups to challenge the state. This thesis recapitulates longstanding orientalist narratives of rural and third world environmental degradation, moving between tragic or romantic depictions of climate-affected Syrians and tropes of unruly masses of migrants overwhelming the cities, radicalizing, and erupting into violence. In the process, the narrative sidelines discussion of the economic grievances and political critiques of the Assad government that led to the 2011 uprisings by broad sections of Syrian society in the midst of the broader Arab uprisings and the Rojava autonomy movement. Furthermore, it fails to place a central focus on the agrarian areas of the Hasakah governorate in northeast Syria, where the effects of the drought were most acute and the development of new experiments in collective land use and ecology challenge narratives suggesting that environmental precarity makes social conflict inevitable.

Reviewing journalism and environmental media that invoke the Syrian climate war thesis, this chapter explores how the framing of the Syrian war and Arab uprisings as being triggered by climate-induced drought helps sustain a neocolonial discourse on both environmental migration and "failed" postcolonial states. Arguing that an overemphasis on both scarcity and Malthusian theories of resource conflict in reporting on the Syrian conflict builds on older, racialized narratives of social breakdown caused by the inability of the colonized to properly manage resources, the Syrian climate war thesis reinforces neoliberal precepts about the purported dependency of rural populations and reproduces a geopolitical mapping of conflict that configures Muslim-majority states as particularly subject to mismanagement and insurgency. In such depictions, representation of Syrians disabled in war helps distill the human tragedy of the violence of climate change. As an icon, the disabled climate refugee evokes the embodied vulnerabilities produced by widely distributed climate changes, in the process masking the socioeconomic processes that contribute to interlinked mass mobility and debility. Although migration is a possible outcome of climate change in the region, an account of diverse narratives of the Syrian war demonstrates that aerial bombing and armed struggles among rebel groups and the state have been more proximate causes of displacement and vulnerability as well as more direct threats to agricultural production. Noting that the climate war thesis has been embraced by some leftist critics of climate change, the chapter argues that a broader accounting of geopolitical and economic relations of migration can help build a more contextual analysis of how environmental factors intersect with the social and political dynamics of war. Finally, the chapter concludes by arguing that climate migration discourses mask understanding of some of the possibilities for different social and ecological futures emerging from revolutionary critiques of state and colonialism in the region.

Environmental War and the Racial Map of Conflict

The growing attention to climate change as a security problem has been influenced by the collapse of the Kyoto regime of international carbon emission controls, which coincided with a shift in emphasis in climate policy circles from mitigation (reducing the use of fossil fuels) to adaptation (developing strategies to manage the effects of higher global temperatures) and from international to national frameworks for action. Such transitions—which echo older traditions of enclosure and Cold War isolationism—are evident in the U.S. approach to climate diplomacy, which has emphasized the potential for global warming to increase transnational armed conflict. In 2007, the U.S. defense think tank CNA famously identified climate change as a "threat multiplier." In the absence of a binding emissions control regime, this phrase has since been adopted by both the Pentagon and the United Nations to suggest policies aimed at adapting to a hotter world rather than collaborating to mitigate anthropogenic warming. The military "threat multiplier" concept invokes neutrality on the politics of climate science, instead focusing on what it considers climate change's undeniable emerging impact. General Chuck Jacoby (Ret.), former commander of U.S. Northern Command, emphasizes the catastrophic potential of resource conflicts in a warming world, which he considers an undeniable effect of climate change even though some Americans choose to deny the cause of those changes: "Many conflicts throughout our history have been based on resource competition. Increasingly in the future, we will be defining our national interest within those resource contests. . . . You can predict that that drives human activity in a way that can create conflict. . . . It can be considered a politicized issue. . . . I deal with the facts. Whatever the cause is less relevant to me than the effect."[5]

Coincident with the rise of "new war" theories that blur traditional divisions of war and peace, civilian and soldier, human and nonhuman, the view of climate change as a threat multiplier reflects a broader attempt by states to bring social, economic, and environmental phenomena under the purview of state security. Yet there is some irony in the specific shift from climate change mitigation to adaptation schemes marked by the rise of environmental security discourse. While there is a scientific consensus on the destructive environmental and economic outcomes of current levels of carbon use, the idea that climate change leads to armed conflict remains one of the more dubious and contested claims among geographers and anthropologists who study environment-society interactions. Yet this idea has become one of the most widely discussed in terms of climate change impact, galvanized by the emergent attention of military planners and international relations scholars to planetary environmental questions. In this nexus of environmental and militarized security thinking, there is a resurgence of environmental determinist framings of human development that recapitulate racial mappings of risk. Projects of U.S. and European imperial intervention have long involved depictions of how heat, scarcity, and underdevelopment render colonized populations less modern. From the nineteenth century into the early twentieth century, climatic determinism focusing on the deleterious effects of heat guided a number of racial and colonial knowledge projects—ranging from behavioralist theories about crime to tropical medical theories of contagion to eugenic notions of race improvement.[6] For Robert Vitalis, these traditions of imperial thought guided the rise of international relations as a field of academic research following World War I. In his book *White World Order, Black Power Politics,* Vitalis argues that the development of the field's first major journal, *Foreign Affairs,* from its predecessor, *Journal of Race Development,* is instructive for understanding how the field's conception of the international was informed by fears of the worldwide decline of white control in the face of anticolonial movements. Vitalis argues further that international relations has witnessed varied rearticulations of its eugenic roots over the decades, not the least in discussions of the geopolitics of oil as a "resource curse" that frames environmental degradation as an intractable problem of underdevelopment in the Global South.[7] The mobilization of race as a mapping device to conceive of the "international" was especially important for the white settler colonies, where race provided a heuristic for evaluating the success of colonial projects in maintaining and extending a broader global project of white development.[8] Racial mapping of threats to the geopolitical dominance of the United States and Europe thus grows out of longer colonial worldmaking projects and offers a naturalized common sense from which security experts worldwide narrate risks arising from purportedly intractable cultural and strategic difference. In representations of international conflict in journalism and policy arenas, it is commonplace for such risks to be written on Black and Brown bodies, returning theories of international conflict from a normative spatial understanding of the international to its racialized, developmental, and environmental-determinist roots.

In its post-9/11 articulation, the racialized geopolitical mapping of climate risk highlights majority-Muslim state collapse and the rise of Islamism as primary threats to the international system. Taking the anticipated environmental displacement of Muslims as a sign of potential radicalization, such invocations of climate-driven risk build on Islamophobic assumptions that have become conventional in post-9/11 writings of Western security experts. Several of the locations that have been designated as climate migration hot spots by climate security experts have been locations where migration is said to exacerbate the threats of Islamist insurgency. These crises are geographically configured within a tropicalizing discourse, where the hotter, wetter, already unstable climatic regions may eventually bring the blowback of anthropogenic warming to the Global North—the regions most historically responsible for anthropogenic carbon emissions that cause today's rapid climate change. Emerging depictions of the figure of the climate refugee reify a racialized first world–third world division that grafts anthropogenic climate change onto a neocolonial racial map of economically dependent former colonial states. In such speculative risk scenarios, the worst-case outcome is articulated as state collapse followed by the rise of Islamist insurgency or an Islamic state. The concept of a "climate refugee" in turn purports to encompass the otherwise hidden ecological contributions to the migration calculus, which may inaugurate a chain of events that break down social defenses against conflict. As explained in chapter 1, climate security discourse makes the case for creating policies to address climate change on the traditional ideological ground of the Right. But such strategies have had little success in producing concerted changes in the energy economy and have instead resulted in the distortion of how environmental factors interrelate with longer-term neoliberal causes of inequality and migration. The only evidence that such strategies work comes from the Far Right, which seizes on any talk of migration to argue for xenophobic border militarization.

If climate security discourse appears as part of a larger Anthropocene discourse that promises that transgressions of the nature–culture binary are a sign of new posthuman politics, the speculative risk scenarios being constructed in climate security circles appear more like a mix of old-fashioned Malthusianism and orientalism. Since the late 1980s, a number of feminist and postcolonial geographers have analyzed how conservation discourses feminize the tropics and render southern environments sites of unruly reproduction, often drawing on earlier tropes of colonial natural history.[9] The narratives of climate-driven impacts on small agriculturalists rehash colonial development narratives that historically highlighted population pressures and rural-to-urban migration as crises of the state, forgoing analysis of how drought and weather events fit into the political economies of capitalism. Feminist development scholar Betsy Hartmann puts it as follows:

> For those familiar with . . . neo-Malthusian models of environmental conflict developed in the 1980s and 1990s, climate refugee and conflict narratives seem very much like old wine in a new bottle. . . . Drawing on old colonial stereotypes of destructive Third World peasants and herders, [these] degradation narratives go something like this: population-pressure induced poverty makes Third World peasants degrade their environments by over-farming or over-grazing marginal lands. The ensuing soil depletion and desertification then lead them to migrate elsewhere as "environmental refugees," either to other ecologically vulnerable rural areas where the vicious cycle is once again set in motion or to cities where they strain scarce resources and become a primary source of political instability.[10]

Degradation narratives play a particular role in academic and nonprofit discourses on the relationship between southern environmental disaster and migration, as degradation is understood to be a deterministic outcome of poverty as well as a source and effect of displacement. In a key essay on the exhaustion of liberal paradigms for studying third world environmental problems, Raymond Bryant argues that degradation narratives make four interconnected interpretive errors: (1) a deterministic view of poverty as inherently leading to destructive environmental practices; (2) a failure to differentiate the global poor by subsistence practices, location, race, gender, or other factors; (3) an emphasis on first world environmental intervention as a presumed solution to degradation, without consideration of the role of first world economic policy in creating poverty; and (4) a naturalization of poverty geographically located in the third world, which subsumes those affected by poverty into developmentalist progress narratives. As such, degradation narratives view the participation of the poor in environmental destruction as inevitable, and the intervention of NGOs and northern states as natural solutions to the deterministic effects of poverty on the environment.[11] Jan Selby and Clemens Hoffman suggest that scholars pay attention to the ways that such deterministic precepts about environmental causes of conflict have come to be embedded as common sense in some academic and policy sectors; narratives such as the Syria climate war thesis are troubling evidence that "most climate security discourse is . . . indebted to the Malthusian tradition for its core ontological and political premises," as it "tends to interpret the global poor, and sub-Saharan Africans in particular, as the most likely subjects—and also sources—of climate related conflict."[12]

The speculations of risk articulated in the Syrian climate war thesis reflect how stories of the tragic lives of climate refugees are coming to embody the hope for militarized environmentalist intervention as climate research and policy becomes suffused with the logics of state security. As the face of capitalism's self-destructive consumption and waste practices, the climate refugee demonstrates how unchecked and rapid warming of the post-Kyoto era is considered by Western environmentalists to be a proximate cause of destabilizing challenges to the international order.

Drought and Disability: The Syrian Climate Refugee

Journalism and policy reports on environmental migration often make speculative causal claims about the relationship of climate change to migration. By focusing on the vulnerability of Syrian refugee bodies, including those of women, children, and the disabled, such reporting attempts to distill widely distributed geophysical changes in climate into icons of climate-driven suffering. In a classic essay on disability representation, David Mitchell and Sharon Snyder note that the disabled body itself becomes ideologically linked to social visions of crisis, as its purported vulnerability provides evidence of twinned problems of representation and sovereignty requiring repair: "The perception of a 'crisis' or a 'special situation' has made disabled people subject not only of governmental policies and social programs but also a primary object of literary representation." The fact that the disabled body can at once operate as a "stock feature" of narrative and "an opportunistic metaphorical device" helps us glimpse the ideological vector of disability as social crisis, for it is in the disabled body that an icon of vulnerability signifies a broader ecology requiring intervention.[13] Embedded within contemporary crisis representation, then, is the sentimental rhetoric of humanitarianism, which Rosemarie Garland-Thomson shows is prevalent in mass-mediated images of disability as tragic loss.[14] This rhetoric was widely circulated in the social media sharing of photos of the dead body of three-year-old Alan Kurdi, who drowned in the Mediterranean on September 2, 2015, when his family attempted to cross from Turkey to the Greek island of Kos. Given the location of such images in war zones and at borders, it appears that disability signifies crises of state power that can be captured by neocolonial discourses of humanitarian rescue, reifying the state as the ultimate arbiter of security. The vulnerable body in war—whether conceived as disabled, childlike, or feminized—reflects a broader logic of colonial representation, justifying intervention of the state in the purportedly patriarchal or underdeveloped conditions that allow for the mass reproduction and exploitation of social vulnerability.[15]

Disability thus plays an important role in the public representation of the "migrant crisis" of the eastern Mediterranean. When attached to representations of national vulnerability, images of disability help to mask the manner in which national borders both create zones of privileged security and reproduce transborder violence. Much of the existing scholarship focusing on this as a media event has emphasized how language about refugees in Europe in 2015–16 shifted from initial humanitarian sympathy to increasingly xenophobic invocations of parasitism, threat, or social conflict.[16] But this is only part of the story; as Black studies scholars focusing on race and sexuality in European contexts have noted, race is more often managed as a permanent crisis of social incompatibility that is used to restrict citizenship and disavow the colonial legacies evident in migration.[17] Disability can furthermore be deployed to suggest that different countries exhibit developmental differences, masking the deeper structural roots of conflict and the specific geopolitical relationships established by militarized interventions of other state and non-state actors. To view the Syrian crisis in the longer context of racial capitalism, it is necessary to emphasize (1) that racial animus against migrants in Europe and white settler colonies recapitulates a history of violence and displacement under border imperialism, and (2) that transnational migration recruits spatial distinctions between nations

Syrian refugees arrive on Lesbos, in Jessica Corbett, "'We Have to Get This Right': Historic Bill in the US House Would Create Specific Protections for Climate Refugees," Common Dreams, October 14, 2019, www.commondreams.org/news/2019/10/24/we-have-get-right-historic-bill-us-house-would-create-specific-protections-climate. Photo: © UNHCR/Andrew McConnell.

toward the racialization of labor.[18] Given this linkage of race and labor through transnational migration, disability allowed Syrian migrants to be branded initially as tragic victims despite the later attribution of criminality or terrorism.

As such, visual depictions of Syrians entering Jordan, Lebanon, or southern Europe displace the geographic configuration of security discourse from Syria itself to other international borders. They provide a focal point for generic reporting on the worldwide potential for climate-driven displacement, resource conflict, and war. An article on the Common Dreams website uses a photo of a Syrian man emerging from the Mediterranean, carrying a child to the Greek island of Lesbos, as an illustration of the need to resettle climate refugees. In the background, a group of refugees climb off an inflatable craft, as another man carries a second child toward the island.[19] Such images of the drenched bodies of Syrians disembarking from perilous Mediterranean journeys became stock features of northern reporting on the potential of the United States and Europe to serve as protectors of Syrians and other refugees. At the same time, such images were ambivalently configured as signs of risk, with precarious refugee bodies bringing with them the potential of terrorism or economic loss. In 2016, *Newsweek* published the article "Should Europe Be Concerned about Climate Refugees?" accompanied by a photo that depicted a large group of Syrian refugees crossing into Jordan at the Hadalat crossing east of Amman. Foregrounding a group of parents struggling to carry children and belongings, headscarves shielding their faces from wind and dust, the article cites the *PNAS* study to claim, "Drought linked to climate change devastated rural areas in Syria, driving people to overcrowded cities and fueling discontent in the urban centers where protests first erupted in 2011." Reiterating stereotyped conflict geographies that separate North and South into zones of stability and violence, the article presents Europe as a "haven" despite the widespread xenophobic animus emerging during the so-called refugee crisis. In the process, authors Rob Bailey and Gemma Green figure war and climate change as the cause of a refugee crisis that European states could have predicted but did not:

> Ongoing conflict in the Middle East and the unsustainable accumulation of refugees in neighboring countries should have been warning enough for Europe's governments. Things are unlikely to improve any time soon. Europe is a haven of stability in a neighborhood of fragility. From North Africa to the Middle East and across the Sahel into the Horn of Africa, a great many of Europe's neighbors are at risk of, or experiencing, conflict. Climate change will make a bad situation worse. As a recent report for the G7 argued, it will undermine livelihoods, increase local resource competition, aggravate pre-existing tensions and destabilize markets, ultimately increasing the risk of social upheaval. In extreme cases, climate change may leave people with little option but to move. One recent analysis found temperatures in the Middle East and North Africa could be so extreme by the end of the century that some areas may become uninhabitable.[20]

Reporting on Syrian migration need not engage with definitive claims about causality in order to use the disabled or vulnerable Syrian body as sensible evidence of climate-driven conflict threats. Uncertainty about the causes of war generates speculative associations. In the British newspaper the *Independent*, publication of a vegetation loss map of northern Syria spurred the question, "Did climate change help spark the Syrian war?"[21] In an instance when a journalist takes account of scholars' caveats about the weak evidence of causality between climate change and the Syria climate war thesis, Phil McKenna nonetheless uses images of a Syrian refugee as the

Should Europe Be Concerned About Climate Refugees?

ROB BAILEY AND GEMMA GREEN
ON 5/18/16 AT 7:20 AM EDT

Syrian refugees carry their belongings as they wait to enter the Jordanian side of the Hadalat border crossing, a military zone east of the capital Amman, after arriving from Syria, May 4.
KHALIL MAZRAAWI/AFP/GETTY IMAGESMAY 4, 2016.

Rob Bailey and Gemma Green, "Should Europe Be Concerned about Climate Refugees?," *Newsweek*, May 18, 2016, www.newsweek.com/should-europe-be-concerned-about-climate-refugees-460661.

face of the future climate migrant. In an interview published by the nonprofit website *Inside Climate News*, Dutch environmental policy professor Frank Biermann emphasizes that even though "many of these refugees come from countries affected by climate change . . . I would not make necessarily any causal link between climate change and the Syrian or Iraqi crises. Of course, there are many other reasons responsible for the war and civil strife in these countries." Nonetheless, the article displays an image of a lone Syrian child in a refugee camp in Lebanon; the caption claims the child "might be foreshadowing bigger refugee crises ahead" due to climate change.[22]

The sources of vulnerability for Syrian refugees in such images and accounts tend to be drought, malnutrition, and the rapid breakdown of social safety. The result according to these articles is the mass debilitation of populations, which results in their increasing mobility. A key report by the British medical journal the *Lancet* makes the connection between climate change, migration, and the breakdown of physical and mental health. Syria becomes an example of such connections in a series of links between climate change, drought, malnutrition, displacement, and war. According to the 2018 edition of the journal's annual report on climate change and public health, "In Syria, many attribute the initial and continued conflict to the rural-to-urban

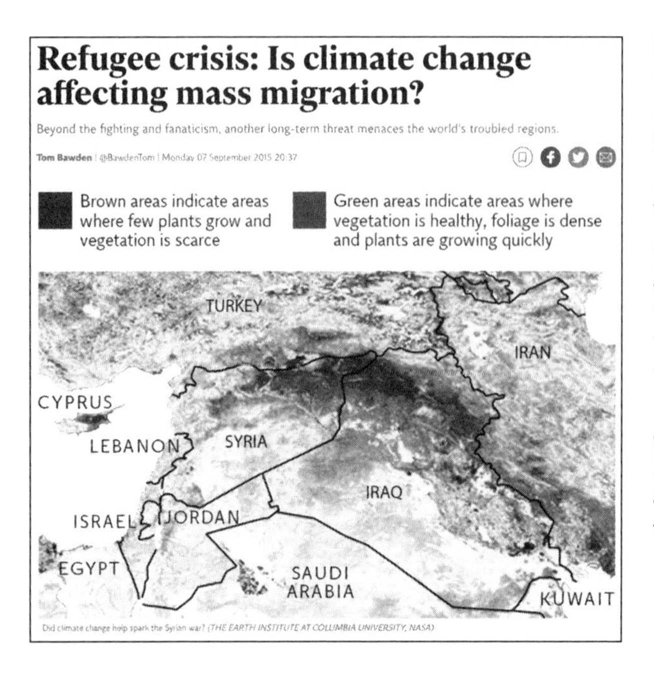

Refugee crisis: Is climate change affecting mass migration?

Beyond the fighting and fanaticism, another long-term threat menaces the world's troubled regions.

Tom Bawden | @BawdenTom | Monday 07 September 2015 20:37

Brown areas indicate areas where few plants grow and vegetation is scarce

Green areas indicate areas where vegetation is healthy, foliage is dense and plants are growing quickly

Did climate change help spark the Syrian war? (*THE EARTH INSTITUTE AT COLUMBIA UNIVERSITY, NASA*)

Tom Bawden, "Refugee Crisis: Is Climate Change Affecting Mass Migration?," *Independent*, September 7, 2015, www.independent.co.uk/news/world/refugee-crisis-is-climate-change-affecting-mass-migration-10490434.html. Lighter gray areas indicate normal foliage, darker gray areas indicate scarce vegetation.

Migrant Crisis: 'If We Don't Stop Climate Change...What We See Right Now Is Just the Beginning'

A Q&A with Frank Biermann, a Dutch researcher who led a controversial 2010 study on climate refugees, who fears crises like Europe's will only get worse.

BY PHIL MCKENNA

SEP 14, 2015

Syrian refugees, like those currently settling in Lebanon, might be foreshadowing bigger refugee crises ahead. Credit: CAFOD, via Flickr.

The surge of people fleeing to Europe from the Middle East highlights how quickly mass migrations can occur. It may also offer a glimpse of what's to come as climate change makes some regions around the world unlivable, according to a leading researcher on the human effects of climate change.

Frank Biermann, a professor of political science and environmental policy

Frank Biermann, "Migrant Crisis: 'If We Don't Stop Climate Change . . . What We See Right Now Is Just the Beginning,'" interview by Phil McKenna, *Inside Climate News*, September 14, 2015, https://insideclimatenews.org/news/13092015/migrant-crisis-syria-europe-climate-change.

migration that resulted from a climate change-induced drought. However, the factors leading to the violence are wide-ranging and complex, with clear quantifiable attribution particularly challenging. Indeed, climate change, as a threat multiplier and an accelerant of instability, is often thought of as important in exacerbating the likelihood of conflict. Nonetheless, migration driven by climate change has potentially severe impacts on mental and physical health, both directly and by disrupting essential health and social services."[23] Here, after the caveat about uncertainty, the authors of the report connect widely distributed climate changes to the embodied vulnerabilities of migrants fleeing conflict. Such representations of disability as an indicator of climate change are increasingly prevalent in northern journalism on international conflicts.

An article on Syrian migrants is an extended case in point. The cover story of the March 2016 issue of *Scientific American* announces that "fugitives from Syria's devastated farmlands represent what threatens to become a worldwide crush of refugees from countries where unstable and repressive governments collapse under pressure from a toxic mix of climate change, unsustainable farming practices, and water mismanagement."[24] In the article, journalist John Wendle argues that "drought, which is being exacerbated by climate change and bad government policies, has forced more than a million Syrian farmers to move to overcrowded cities."[25] Despite the large numbers of farmers portrayed as climate refugees, the story and accompanying photo essay offer one story of a farmer fleeing Aleppo and another of a formerly successful well-digger who left the outskirts of Kobane. These stories are accompanied by a general series of photographs depicting the arrival and settlement of refugees on the Greek island of Lesbos, where Wendle conducted interviews.

For most of its text, the article relies on the research by Kelley and colleagues in *PNAS*. Notably, that study does not analyze the local hydrological effects of drought within Syria, instead making arguments about the broader Fertile Crescent region. The drought in Syria was most severe in the desert east of the country, and its epicenter was the northeastern Hasakah governorate, which falls in the Kurdish-majority Rojava autonomous region (and which at the time of this writing was being occupied by a combination of Turkish and Syrian Islamist forces). A detailed review of the *PNAS* article and broader claims of the Syrian war being driven by climate change was published in *Political Geography* in 2017 by Jan Selby, Omar Dahi, Christiane Fröhlich, and Mike Hulme, who highlight the significant effects that the Assad government's neoliberal agricultural reforms had on small farmers in the lead-up to the drought. They argue that "inattention to Syria's changing political economy" led proponents of the Syrian climate war thesis "to systematically overstate the impacts, both direct and indirect, of the 2006/7 to 2008/9 drought on migration."[26] Paying attention to the specific types of water use for the main agricultural sectors of different parts of the country, the authors conclude that the drought's most lasting impacts on agriculture in the East were not ultimately coincident with the large-scale migration events described in the Syria climate war thesis. This conclusion is backed up by later research looking at the effects of the drought across Turkey, Iraq, and Syria, which stresses the relationship of the state to infrastructure as one of the key components affecting yield outcomes.[27]

Neoliberal policies in the later years of the Bashar al-Assad government are more likely reasons that the breadbasket of northeastern Syria transformed from an agricultural exporting region to a region in which farmers had persistent difficulty accessing infrastructure to maintain operations through the regular cycles of drought. Although divisions between opposition groups posed challenges for developing solidarities against the Assad regime, the emergent Syrian uprisings in many ways echoed grievances against Assad that were articulated following state attacks against Kurds in northeastern Syria in 2004. Nonetheless, it was only after the onset of the broader revolution in Syria that the Kurdish-led forces were able to more formally declare autonomy from the state. Analyzing a small number of interviews with refugees in Jordan, the *Political Geography* article is perhaps too quick to dismiss the idea that migrants from northeastern Syria may have been politically involved in the uprisings; although the Kurdish and Arab uprisings were conventionally depicted as separate political causes mapped onto a majority-minority divide, Hasakah city was the site of some of the first anti-regime protests of 2011, led by Kurds, and the retrospective interviews in refugee camps may be affected by the concern for retribution among migrants. Still, the broader points of the authors that political and economic grievances were deep drivers of the war and that there is insufficient evidence that climate change was a substantial trigger of migration in this area are sound.

It is beyond the scope of Wendle's article in *Scientific American* to substantiate his claim, taken from the *PNAS* article and repeated widely in the news media, that "as many as 1.5 million Syrian farmers" were displaced by the drought. But Wendle configures Europe as a place of rescue in the face of this scale of mass migration. Displaying images of a man crying on the side of a road and a woman in a headscarf clutching a baby at Pikpa, the

Photo of man kneeling, in John Wendle, "Syria's Climate Refugees," *Scientific American*, March 2016, 52.

Photo of mother and child, in John Wendle, "Syria's Climate Refugees," *Scientific American*, March 2016, 52.

caption suggests that these experiences reflect the generalized feeling among migrants that they are "overwhelmed with relief upon reaching the Greek island of Lesbos." Alongside this broader conception of Europe as a haven, Wendle employs a disability narrative in order to conceptualize the human cost of drought. He focuses on the individual story of fifty-four-year-old well-digger Kemal Ali to illustrate the association of climate change and migrant crisis. Describing how during the 2006–10 drought, the depth of wells Ali drilled grew from seventy to seven hundred meters, Wendle links a series of economic, social, and physical factors to the trigger event of a declining water table: "Ali's business disappeared. He tried to find work but could not. Social uprisings in the country began to escalate. He was almost killed by crossfire. Now Ali sits in a wheelchair at a camp for wounded and ill refugees on the Greek island of Lesbos."[28] Representation of disability in the tragic mode helps Wendle account for the force of climate in unleashing a chain of disruptions in Ali's social world, resulting in his family's eventual passage to Turkey and then, by boat, to Greece:

> Ali likewise tried to stick it out, but few of his former customers could afford to drill as deep as the water had sunk. And the war made ordinary activities practically impossible. His home village was only a short distance from the wreckage of Kobane on the Turkish border. That town was in ruins by the time the Kurds succeeded in recapturing it from ISIS, the militant group that has been terrorizing the region. Last July he headed for Syria's capital, Damascus, hoping to find work and a place where his family could be safe. He was on his way there by bus when a rocket struck the vehicle. He awoke in a Damascus hospital, paralyzed from the waist down. The blast had peppered his spine with shrapnel. Somehow his family managed to get him back north, and together they made their way across Turkey to the shores of the Aegean.[29]

Ali's paralysis is the culmination of a series of factors—economic hardship, the arrival of ISIS, and the dangerous urban-to-rural migration—purportedly triggered by drought. Disability here serves as a means of distilling the widely distributed effects of the climate system on affected migrants, as those fleeing war conjoin the specter of mass debilitation under climate change with the intensification of global mobilities wrought by ecosystemic collapse. Uniting disabled bodies with hypermobility under conditions of war, such narratives attempt to anthropomorphize the climate system in the bodies of vulnerable Syrian migrants—figures like Ali, the children emerging from the waters of the Mediterranean, and the women struggling to cross the Jordanian border.

Although the narrative of Ali's transformation from successful laborer to paralyzed refugee begins with drought, it is notable that Ali's actual decision to leave his home came not with the depletion of his business but later, in July 2015, following the nearly yearlong siege of Kobane and surrounding villages by ISIS and the June 2015 Kobane massacre, in which ISIS carried out a series of suicide attacks. Control was reestablished by the YPG/YPJ units of the Syrian Democratic Forces following a U.S.-led air bombardment that devastated the city. None of these events—which represent more sudden and severe forces of displacement—appear in Wendle's article. Instead, Wendle emphasizes issues of water and agricultural collapse in producing disability. The article displays a blurry photograph (not reproduced here) of a resting Ali attended by his family members at the Pikpa camp on Lesbos alongside the following narrative:

> Ali and his family are trying to somehow get him to Germany, where they hope surgeons will be able to restore his ability to walk. Outdoors in his chair to get a few minutes of sun, Ali is thinking of the friends he left behind in Syria. "The life of a farmer has always been hard," he says. "Their biggest problem was water—period. Because water is life." His son wheels him indoors for a rest. Weak winter sunlight partially illuminates a big room lined with a couple of dozen beds. Plastic sacks and cheap duffle bags are heaped everywhere, holding their owners' few remaining possessions. As Ali's children lift him into bed, his face crumples in pain and exhaustion. Fardous, his 19-year-old daughter, tucks his colostomy bag against his body and arranges the donated blankets to cover him. "It is written in the Quran," Ali repeats. "Water is life."[30]

Wendle ends the article here, invoking both the Quran and the phrase "water is life," which has become an Indigenous rallying cry against oil pipelines since Standing Rock activists popularized its translation from the Lakota *mni waconi*. Although Ali remains paralyzed, hope for asylum is presented as a path forward for the family, configuring impairment as the tragedy that can be overcome in the act of resettlement. Whereas this overcoming of disability does not take place in the dramatic acts of adaptation attributed to the char island dwellers depicted in chapter 3, Wendle's article situates the refugee journey as a passage away from the persistent threat of death and toward a new horizon of life.

Islam is obliquely referenced as the backdrop of the refugee passage. Inscribing a Quranic invocation of protest against the carbon economy, the end of the article plays on a contrast between an environmental ethic and the association of Syria with the rise of ISIS and Islamist insurgent groups. Although it is only implied in Ali's story, international security discourse on the Syria war as a climate war draws on the specter of ISIS as climate change's ultimate challenge to the liberal secular state system. Elsewhere in the climate security literature, the association is more explicit. In a study commissioned by the German government, the German think tank Adelphi contends that drought led directly to both the onset of war and the rise of ISIS and the al-Nusra Front. The report relies heavily on the listing of ethnic groups—ranging from Alawites to Kurds to Sunni Arabs—as a shorthand for conflicting interests already present in the region. In this already potent mix of social differences framed by poor state governance, Adelphi claims that climate change, "though far from being the only or the primary driver of conflict in Syria, . . . did play a catalytic role in accelerating the descent into fragility and facilitating the rise of NSAGs [non-state armed groups]." Emphasizing resource conflict, the report claims that ISIS was able to exploit water scarcity as a recruitment tool for migrant herders:

> Farmers and herders in the northeast who were faced with crop failure and livestock death had little to no economic prospects and there were no adequate social safety nets in place for them under the Assad regime. As ISIS pays its fighters an estimated USD 400 per month, about five times as much as a normal wage in the region, it also provides economic incentives for young and unemployed people with few perspectives [*sic*]. Economic hardship is a primary driver for Syrians to join armed groups, as unemployment reaches up to 90 percent and most salaries of those who still have employment are insufficient for meeting basic needs.[31]

From Climate War to Climate Revolution

In her memoir *The Crossing*, journalist Samar Yazbek frames her memories of the Syrian war as a process of dismemberment occurring on parched earth: "In my mind, I hold a portrait of Syria, but it is no ordinary image. It shows a dismembered collection of body parts, the head missing and the right arm dangling precariously. Then you notice a few drops of blood slowly dripping from the frame, disappearing as they are absorbed by the dusty soil below. This is the catastrophe Syrians deal with every day." Noting the persistent "signs of drought" that left "hardly any signs of life in the landscape," Yazbek's descriptions of a series of journeys across the western side of the border separating Syria and Turkey in 2012 and 2013 represent wartime Syria as a place where olive trees shrivel, villagers struggle to eke a living from tending starving sheep, and people shelter indoors, fearing the fate of those who lost limbs, life, and loved ones in the crossfire.[32] Portraying her visit to a Turkish hospital at Reyhanli, where she describes "putrefying" bodies "laid out on white sheets, with mutilated feet, amputated limbs, hazy eyes," Yazbek uses the images of bodily fragmentation to metaphorically portray the dissolution of the society, described as the "shattered heart" of Syria in the book's subtitle.[33] The only rain that appears in the narrative is the raining down of shells on the border region. Yazbek describes the aerial bombing and sniper shelling carried out by Syrian, U.S., and Turkish forces as an atmospheric violence, representing a "blazing sky," a "sky that wouldn't let us celebrate; no, the sky was on fire."[34]

For Yazbek and some other memoirists of the Syrian uprisings and civil war, this atmospheric violence is not a violence primarily produced by the drought and other climate-driven processes. Aerial bombing and ground battles that intensified with the increased number of armed fighters and the growing number of regional parties to the conflict are depicted as generating both an intensive toll on the national body and a breakdown of social and ecological conditions for reproducing life. Although such depictions of dismemberment are common to war memoirs in a variety of historical and geographic contexts, the connection between the collapse of social reproduction, portrayals of aerial bombing, and visions of the bodily and architectural fragmentation in war have framed the context of narratives of war and migration in Syria. In Marwan Hisham's *Brothers of the Gun*, the winter food shortages in Raqqa are depicted as an effect of the intensifying battles over infrastructure emerging as the disparate rebels—including the groups that would re-form as ISIS—took over the grain silos in the east. As he waits in a long line outside a bakery in December 2012, Hisham portrays his internal struggle to make sense of how the twin violences of the state and the rebels, "prison and anarchy," intertwine to deprive people of food and energy in Raqqa, along with the neighboring breadbasket provinces of Deir ez-Zor and Hasakah: "Surely, this is temporary. It will end when the rebels come. It's a siege situation. The rebels are going to break through fast and lift it. They care what we think about them. But do they? Aren't they the reason I'm standing here? Aren't they the reason those people are behaving

Marwan Hisham, *Brothers of the Gun: A Memoir of the Syrian War*, illus. Molly Crabapple (New York: One World, 2018), 182. Courtesy of Molly Crabapple.

like animals? Or not? Some say that the Department of Subsidies is making a profit from this situation and also that the government officials want us to hate the rebels in advance."[35] Whereas Yazbek's literary prose abstracts Assad's aerial bombing campaigns as fire in the sky, Hisham depicts the monstrosity of the planes—which shot "flames from the dragon's mouth"—and emphasizes the personal agency of the pilots, who, disrupting the "calm" of Hisham's "garden," "killed everything beautiful in this country."[36] Although Hisham focuses less on sensational images of bodily trauma, his narrative of the war—from the blood of martyrs at Raqqa's clock tower, to ISIS's public executions, to the U.S. aerial bombing campaigns, to the moment of his exit across the Turkish border near Afrin—focuses on the moments when buildings collapse and bullets or shattering glass pierce the skin, leading masses of people to flee either to the hospital or away from the fighting. The images accompanying the narrative drawn by artist Molly Crabapple capture the connection between aerial bombing, infrastructural vulnerability, and bodily debility; in one image, the arrival of U.S. bombing campaigns in the east produces a landscape of dead bodies that must be extracted from the rubble. In such narratives and images, the vulnerability of bodies to war environments moves from longer time scales—witnessing the slow withering of the population broken down by war—to shorter ones, as sudden explosions generate transformation of the social landscape through sudden migrations.

Given this alternative vision of war as an embodied socio-ecological cataclysm, it is possible to critically analyze how some refugee and analyst narratives of the Syrian war have woven images of embodied and ecological precarity into the climate war thesis. As a primary example, videos in the Weather Channel's *Climate 25* project claim that drought led directly to the Syrian revolution. In one video, Thomas Friedman, the *New York Times* opinion columnist who elsewhere refers to himself as a "tourist with an attitude," claims that "a million Syrians—farmers and herders, really conservative people—left their ranches and fields in the countryside and swamped all the major Syrian towns and cities as they were really driven off the land by the drought, putting huge pressure on the infrastructure. These farmers did not start the revolution, but with the first call of 'Allahu Akbar' by the revolutionaries, they were very eager to join to bring Assad down. There were many, many climate refugees among them."[37] The *Climate 25* series includes related testimony from a Syrian informant emphasizing a chain of events leading from drought to migration to Islamic revolution. Syrian refugee and New America Foundation fellow Farah Nasif gives a short testimonial concluding that "everything changed" with the drought event, provoking public outrage at the government and the eventual rise of Islamism. In the video on the *Climate 25* website, accompanied by links to media reports of the *PNAS* study, Nasif concludes that drought generated the individual feeling of anger that led many in the society to reject Assad. Displaying a framed photo of her family's agricultural land, which shows fallow fields and a small group of sheep and goats grazing, Nasif testifies to the power of the drought to create hostility toward the Assad government:

> We were OK. . . . Like any normal family. Everything changed when the drought [came]. . . . Everyone suffered from those sandstorms. This is our land. All lands become like this—completely yellow desert. Nothing, no life, no chance to do anything. . . . Government doesn't

> help us at that time in anything. The drought is one of the main reasons for the revolution. They have that, ugh, that angry, that annoying, that hate for the government, for the Assad government and what he do for them. They said oh the government doesn't help me before and I don't expect any future, so I will destroy it. Last year I leave Syria forever because the situation is become very horrible. The Islamists become more power, they targeted the women, they targeted the activists, they threaten me, threaten my family, so I was not even able to return back to that part of Syria.[38]

There is tension here in the framing of the revolution as both popular and determined by climate. Notably, although the drought arose in 2006, Nasif's family, like Ali's, actually migrates with the intensification of conflict, especially as ISIS gained ground in the east in 2013. Sources of violence, such as the actions of regional powers and the U.S. arming of Islamist militias, do not appear in the narrative. This video as such reflects some of the ironies and ambivalences of grafting Syrian opposition narratives onto the climate war thesis. Taken together, the videos of Friedman and Nasif suggest that climate change spurred instability and dissent in Syria, yet the ultimate lesson about whether a revolution in Syria could be positive remains unarticulated. The title of Nasif's video—"How Can Climate Change Help Ignite a Revolution?"—could be interpreted to suggest that such a revolution against Assad's dictatorship was necessary, but Friedman's testimony invokes stereotyped visions of Islamist extremism that reinforce military security narratives of climate change as a trigger event for war.

Such stories about drought and revolution in Syria did not come out of thin air. They are the most common types of narrative that make claims for climate change as a trigger of war and political transformation. In 2007, a number of researchers, journalists, and policy makers, including UN Secretary General Ban Ki-moon, designated the war in Darfur as the world's first drought-intensified climate conflict. Subsequent studies have questioned this designation, arguing that the timing of the drought was not coincident with the outbreak of armed violence in Sudan. However, even among some commentators who are skeptical of exaggerated narratives of climate war, rhetorical gestures toward the influence of climate change on revolution sometimes guide speculations about conflict triggers. Parenti's book *Tropic of Chaos* begins with a theoretical model that discusses today's "catastrophic convergence" of poverty, militarism, and environmental change but at times amplifies claims about the triggering force of climate change, particularly in the rendering of militant Islamism and ethnoreligious violence more broadly.[39] Parenti depicts the history of recent wars in Afghanistan as such: "Yes, religious fanaticism, ethnic hatreds, and imperial ambitions are the larger moving pieces, but climate change also fuels the conflict in Afghanistan. First, the violence began as the result of a drought forty years ago. Second, climate stress creates poverty and desperation, which now feeds the insurgency against NATO occupation. Third, climate change causes interstate rivalries, which play out as covert operations inside Afghanistan. Finally, and very importantly, opium poppy is drought resistant to an extent alternative crops are not, and NATO attacks poppy while the Taliban defends it."[40] Despite the book's lack of any discussion of how atmospheric carbon levels in 1970 might have affected drought in the Afghan-Pakistan borderlands, Parenti claims that climate-induced drought contributed to Mohammed Daud Khan's coup against the Afghan monarchy, minimizing the importance of the ongoing set of Cold War political struggles between Communists and Islamists that were taking shape in Kabul.

The boldest set of claims about the relation of climate change to revolution comes from geographer Andreas Malm, who attributes the Egyptian uprising and the Syrian war to climate change. Malm depicts climate change as a potential trigger—though not an underlying cause—that ignites a larger social ecology of neoliberal crisis: "Global warming cannot be a sufficient cause for a revolution: but it can be one ingredient in a powder keg, and it can, at least potentially, light the fuse."[41] Malm's contention about Egypt depends on a series of unsubstantiated inferences about the influence of climate change on world wheat markets, which experienced price spikes leading up to the Arab uprisings: "The bodily metabolism of the Egyptian population revolves around bread, accounting for a full third of daily caloric intake; subsidized bread made of imported wheat is the staple good of the nation. But due to the drought, Egypt received 40 percent less Russian wheat in the second half of 2010 over the equivalent period of 2009. . . . Expensive bread was an aggravating factor, one among several triggers unleashing oceans of dammed up discontent onto the streets, and in this respect, Tunisia and Egypt followed a script as old as revolution itself." Malm goes on to quote Trotsky's description of women textile workers striking for bread in Petrograd in February 1917, which he describes as being "the moment that set the whole train of 1917 in motion."[42] The same goes for Syria, according to Malm. In contrast to Hisham's representation of the bread line as a direct effect of armed siege, Malm focuses on long-term environmental processes transitioning into rapid crisis. Citing the *PNAS* study about the drought, Malm claims, "Estimates range between one and two million displaced farmers and herders. Fleeing the wastelands, they hunkered down on the outskirts of Damascus, Aleppo, Homs, Hama, joining the ranks of proletarians seeking to find a living from construction work, taxi-driving, or any other, mostly unavailable, job. But they were not alone in feeling the heat. Due to the drought, the marketplaces of the country exhibited one of the central vectors of climatic influence on popular livelihoods: doubling, tripling, uncontrollably spiking food prices."[43] Emphasizing the rural roots of the Syrian uprising, Malm repeats some of the conventions of Malthusian development narratives, stressing the instability generated by rural-to-urban migration and land degradation. The fact that Malm reverses the moral opprobrium against impoverished rural migrants in order to idealize their purported revolutionary zeal does not mitigate the distorting effect of such claims about climate as a trigger of revolution.

Given that Parenti and Malm both attempt to synthesize critiques of neoliberal capitalism with accounts of climate change—in the process countering a species-level discourse about human transformation of the environment in the Anthropocene with accounts of class and national differences among humans—it is notable that they emerge as proponents of the language of trigger events, which frames the "threat multiplier" rhetoric of the military establishment's emerging climate security discourse. Such speculative renderings of climate change as security crisis are intended, in Parenti's and Malm's works, to provide an intersection site for integrating political, economic, social, and ecological analysis. To be fair, they each provide caveats about the complex interrelation of social and ecological factors in conflict, and Parenti repeatedly raises problems in determining causality in the chain of events linking climatic to social processes. However, given the scale at which the accounts are framed, and the speculative cast of their conclusions about drought, such works reflect an embrace of climate change as a trigger of conflicts and displacements. Although they do not embrace the most destructive Malthusian precepts of the climate war thesis, their particular attempts to shift from an "old materialism" (one based on a notion of the human capacity to labor and transform nature) to a "new materialism" (one that stresses the agential power of nature itself) smuggles in elements of climate determinism.

Other emerging strands of materialist analysis—particularly among the works of geographers working in the tradition of political ecology—offer more promising pathways for understanding the interrelation between climatic processes and geopolitical economy. Jesse Ribot, a geographer who is studying the causes of migration for Senegalese farmers leaving for Europe, argues that exclusion from markets and political processes are the proximate causes of precarity in many climate-affected regions of the Sahel. For Ribot, writing on the abstractions of Anthropocene discourse, the underlying material bases of vulnerability cannot be understood by centering the peasant-environment relation as in Malthusian degradation narratives. Instead, it is necessary to highlight a chain of factors that affect peasant households' capacities for resilience in the face of environmental change:

> Grounded social-science research does not explain the precarity of the peasant household or its security and ability to withdraw into subsistence as a mere proximate relation between a household and the environment or hazard. Precarity and security are explained by locating the individual in the household, community, polity, market, nation and a differentiated global political economy. They are explained by people's political leverage to shape these contexts. This applies to any social analysis of precarity—of the peasant, the young, the old, the disenfranchised—including climate-related vulnerability analysis. In the Anthropocene, some causal analysis must trace stressors to greenhouse gas effluents, explaining how these effluents are enabled and how their regulation and mitigation are products of a complex social and political-economic history. These are the causes of stressors in the sky. They are distinct from underlying vulnerability.[44]

Beyond Determinism

One issue that arises with international representations of Syrian migrants is that the linkage of embodied vulnerability to state power can configure the Assad government as the ultimate arbiter of security. However, when writing about the geopolitics of migration from Syria, northern journalists and policy experts generally avoid reflection on the relationship of the underlying statism of security discourse to critical discourse on Syrian sovereignty, setting aside discussion of the multiple histories of political opposition to Assad and the broader challenges such opposition offers to international systems of state and capital. Between the aerial bombing, urban sieges, and chemical attacks of the Assad government and the exclusionary violence of ISIS and other Islamist segments of the opposition, international journalists focusing on the Syrian war tend to postpone the ideological questions raised by activists in the revolution—questions that must be at the forefront of analysis of the ongoing war with Turkey's latest incursion into Rojava, which at the time of this writing had added an additional three hundred thousand displaced people to the staggering eleven million who had already fled from homes within Syria's prewar borders.[45] This rhetorical outcome in which journalists skirt the political complexities of the revolution is one that has been persistently critiqued by Syrian leftists, who argue that northern silence in response to calls for solidarity with the Syrian opposition is an extension of the colonial formulation of a mandate to intervene, following French and British strategies for establishing authority through managed ethnic social stratification. Exiled communist dissident Yassin al-Haj Saleh explains:

> The premise itself deems the Syrian revolution to be essentially majoritarian and anti-minority, without a clear explanation for why that is the case, and without showing sensitivity to time and historical changes in the course of the last eight years and for decades before. The roots of Western "neutrality" toward the Syrian revolution are based in this premise. Most Westerners are repelled by the Syrian regime, but they are equally or even more repelled by the Islamic core of our societies. . . . The emphasis placed on "the protection of minorities" is a vocal implication of this amoral neutrality, which is essentially apathy.[46]

For Saleh, the specter of Islamism, which has convinced outside observers to abandon the opposition, has been effectively cultivated by Assad and his allies in favor of a "culturalist" approach that emphasizes modernization and capitalist development along colonial lines. From this perspective, both narratives about the Islamic character of Syrian society and geopolitical narratives about Syria as an example of geopolitical insecurity performatively "depopulate" accounts of Syria's civil war. Saleh explicitly rejects the Syrian climate war thesis as part and parcel of an erasure of Syrian knowledge and politics and a devaluation of Syrian lives:

> The dominant discourses that share the act of producing knowledge about Syria, Palestine, Iraq, and the Middle East—the Geopolitical discourse and the culturalist one—are both depopulated, reductionist discourses that helped greatly in making local populations invisible, indeed nonexistent. These discourses have a dehumanizing effect that

> made our deaths something unimportant. The other face of this invisibility is the disproportionate visibility of factoids related to religion, sect, and ethnicity: every mediocre Middle Eastern "expert" knows that so-and-so is an Alawi, so-and-so is a Christian, or a Kurd. The "rest" are the "majority" Arabs and Muslims that the West should take great care to protect the minorities from its primordial threat. That is why the coverage of Syria and the attitudes of the right wing and left wing media in the West were really scandalous. . . . And there prospers in the United States a theory of explaining our struggle through drought! Four years of drought preceded the revolution and caused it. So it's not a matter of politics, or of social demands or of a thuggish ruling junta. It's not what those irrational Syrians think; science says it is . . . drought. But this science is full of politics as much as it suffers from ethical drought. This environmentalist approach could be fully embraced by neurotic thugs like Bashar Assad, the same way he embraced the culturalist theory that absolved him of the horrible crimes his regime committed.[47]

Narratives of climate migration and climate war in Syria avoid confronting the particular struggles within the opposition—which vary widely, for example, between the various factions fighting for control of the eastern oil fields near Raqqa and the ecofeminist anarchism of the Rojava autonomous region farther north. Leaving an ambivalent view of state power within Syria, such narratives treat disability as an icon of insecurity generalizable beyond Syria, threatening the ability of states elsewhere to contain emerging risks to international order.

Despite the persistence of the Syrian climate war thesis, actual attempts to respond to the conjoined situation of war and drought in northeastern Syria have invoked the potential for solidarity as the basis of a different relationship to land and subsistence. This follows the turn, in the last two decades, of Kurdistan Workers' Party (PKK) political prisoner Abdullah Öcalan from the party's original Marxist-Leninist ideology toward social ecology, inspired by the writings of anarchist theorist Murray Bookchin. With the movement's success at establishing limited forms of autonomy, this strategy has been effective in developing local institutions that are able to challenge both capitalist development strategies and authoritarian forms of control coming from the central governments of the region. Although much of the international reporting on this aspect of the Rojava movement has focused on its articulation of a gender critique, evident in women's prominent roles in both local collectives and YPJ armed units, the texts of the movement connect feminist critique with an autonomist vision of ecological resurgence. Given these dynamics within the movement, autonomous collectives in the region during the war have undertaken ecofeminist education projects, reforestation initiatives, and agricultural cooperatives based not on neo-Malthusian precepts of social collapse but on a vision of ecofeminist democracy grounded in anarchist traditions of mutual aid and critiques of social hierarchy. Öcalan directly connects anticapitalist critique to the "cause of women's freedom and nature's salvation," which can be realized in the development of localized and autonomous democratic collectives based in municipalities, as well as rural cooperatives.[48]

Beginning in 2012, the development of over fifty autonomous agricultural cooperatives in Rojava's eastern Cizîrê Canton, formerly in the Hasakah Governorate of northeast Syria, attempted to establish modes of ecologically sustainable social provision. Responding to the social ecologies of war—disruptions to basic provisions and infrastructure, theft of sheep, munitions pollution, toxic smoke from oil fires—participants in the Rojava campaign also developed a broader critique of capitalist agriculture. While much of the academic debate over whether rural-to-urban migration in Syria in the lead-up to the war was a result of drought-fueled agricultural collapse in the breadbasket region, the analysis of social ecology developed by the eastern Rojava activists questioned the capitalist logics of grain production, which during the last two decades of Assad's rule has been both pesticide intensive and monoculture in nature, helping to intensify the problems of deforestation and desertification that threaten long-term food security. For this reason, the communes attempted to supplement livestock and grain production with reforestation efforts and the cultivation of olives and fruit.[49] At the time of this writing, such efforts have been threatened by Turkey; on October 9, 2019, when Turkey launched an invasion, the international commune focused on ecology in the region published on its blog a drawing of a Turkish warplane unloading a bomb over the emerging green bloom of the cooperatives, framed by an explosion and a tank in the background. The image conveys the precarity of such efforts, which until now have suggested that even in the heart of the drought region that the Syrian climate war thesis views as the epicenter of social breakdown and outmigration, efforts to build new forms of social inclusion and ecological sustainability portend futures beyond any deterministic interpretation of how environment influences social relations.

"Facing the War in Northern Syria: A Statement from Make Rojava Green Again," October 9, 2019, https://makerojavagreenagain.org/2019/10/09/facing-the-war-in-northern-syria-a-statement-from-make-rojava-green-again.

Conclusions

The neo-Malthusian discourse of climate migration emerges at a time when the rise of right-wing political coalitions and border militarization already conceive of the world map as a place rife with racialized risk. For states that are settler colonies or former European colonial powers, the specter of increased migration appears to generate both representational strategies and governing technologies that aim to manage and contain climate disasters that are increasingly configured as inevitable in the post-Kyoto order. Rather than inaugurating a new materialism, a new dawn of understanding human imbrication in the geophysical forces of Earth, they rehearse the oldest form of materialism, which emphasizes how formal causes in the guise of nature structure social life. In the process, they dispense with close attention to how migration is embedded in the forces of racial capitalism's oil-fueled militarization; the figure of the climate refugee and the specter of the climate war instead offer a tragic disabled icon of crisis that maps cartographies of risk in the Global South, particularly in Muslim-majority states. The Syrian climate war thesis reflects these developments in climate change discourse, signaling that migration will appear as the blowback of the climate system rather than a product of geopolitical struggles in the wake of U.S. wars, the subsequent rise of ISIS, and intervention by other regional and international powers. In response to this discourse, intellectuals and activists from the region—across different leftist segments of the uprisings—have rearticulated their critiques of the Assad government, other regional powers, and non-state militias, arguing for an understanding of how warfare and underlying political and economic grievances condition the forms of migration that have affected a full half of Syria's population since the onset of the war. In the process, they have developed accounts of the war and experiments in collective agricultural management that move away from the spectacle of disability as a national tragedy and envision new social ecologies that depart from a framework of persistent competition and conflict.

Conclusion

At the beginning of this book, I discussed how a 2018 report on the journey of a Honduran migrant caravan heading toward the U.S.-Mexico border was depicted by the *Guardian* newspaper as an episode of climate migration. Relying on expert opinion that the 2015 drought had caused a food crisis, the report related one quote from a member of the caravan describing crop failure following the drought. Although a variety of other parts of the story of the caravans might have been emphasized—including the recent U.S.-supported coup, paramilitary violence, a history of poverty related to international debt, border militarization, and the work of activist groups to organize the caravan—journalists struggle to synthesize stories of such events in ways that are attentive to the complexities underpinning migration. For this reason, a number of online platforms have emerged to shift public discussions of migration and center refugee writings in the critiques of border imperialism and racial capitalism.

In 2019, Migrant Roots Media was founded by activist Roxana Bendezú in North Carolina to add resolution and complexity to public discussions of transnational migration, in part by helping to advertise how receiving countries influence migration flows. One of the website's early publications was a *testimonio* by Alejandra Mejía, who describes her family's journey from Honduras to Panama at the time of Hurricane Mitch, which caused massive flooding and destruction in Central America in 1998.[1] Noting that poverty had already limited economic opportunities in her birth city of Tegucigalpa prior to the hurricane, Mejía's narrative integrates environmental issues related to flooding with a broader context of debt dependency, poverty, and urban violence that converge to promote migration. By reconstructing a life narrative in the wake of violence or displacement, the *testimonio* genre allows for integration of environmental factors leading to migration that might otherwise escape attempts by journalists and policy makers.

The fact that the flooding damage wrought by Hurricane Mitch—which intensified in the storm's aftermath due to corporate logging—predated contemporary journalism on climate migration suggests that environmental problems also have a history that is irreducible to the narration of climate change as a crisis trigger event. Such histories are deeply intertwined with forms of development and land arrangements that make agrarian populations more susceptible to events like Hurricane Mitch or the 2015 drought mentioned by the *Guardian*. One lesson of what is missed in the gap between the *Guardian* article and Mejía's story is that postcolonial agrarian struggles for a share of the wealth generated by extractive enterprise are commonplace worldwide but structured by particular ecologies and social forces that affect people's lives in context. Such struggles can be rebranded generically as the struggles of "climate migration" in environmental media focusing on locations from Kiribati to Bangladesh to Syria to Honduras. This fact reflects not just that public knowledge of environmental problems has been compromised and slow to emerge but also that the social movements challenging environmental destruction may benefit from a clearer articulation of how structural forces of oil-fueled development and capitalist expansion make rural-to-urban labor migration a requirement for large minoritized populations worldwide. The history of attempts to represent "climate migration" as a public problem is a reflection that rural lifeworlds themselves are one of today's targeted geographic horizons of accumulation, rendered commoditizable by both the forces of climate change and extractivist industries that work in tandem to render growing swaths of humanity as surplus labor.[2]

As these processes expand transborder migration, the inequalities between zones of securitization—Christian Parenti's armed lifeboats—and zones of shrinking opportunity intensify impacts on those most affected by the social ecologies combining debt, loss of livelihood, and extractive displacement. One of the difficult challenges that arises in this context is to critically address the fears of mass social breakdown that energize ethnoclass securitization and determinist claims about how climate change might inherently promote social conflict and, thus, racism. This book has argued that climate migration narratives in public media and policy studies have a tendency to mask complex structural forces underpinning migration. Instead of embracing a planet-scale story that contemporary mass displacement is being generated by climate change, I have turned to the recent past of the rise of the oil economy and neoliberal structural adjustment to understand how the world's most rapid processes of industrialization and urbanization—concentrated across Asia—produce racially marginal migrant populations whose vulnerability may be accelerated by environmental forces but whose histories are irreducible to the weather. If in Bangladesh the anticipation of climate-driven extinction of local lifeways underwrites state and NGO discourses of adaptation that preemptively displace agrarian populations, in Syria the climate war thesis works to erase some of the complexities of agrarian challenges that preexisted the war and reflected the broader political discourses of the opposition. In both scenarios, climate migration discourse produces obstacles to allowing environmental movements to integrate histories of war and displacement into narratives of environmental racism, which often retain an ex post facto valence. Rather than alternatively turning to paradigms of environmental harm under the banner of the Anthropocene and its cognate "cenes" (Plantationocene, Capitalocene, Eurocene), these contexts call for a focus on the agrarian base, Indigenous struggles, conflicts over extraction and debt, and the role of the state in creating conditions for migration in national and transnational contexts.

One goal of this book has thus been to consider how it might be possible to link up local particularities with accounts of global processes, with the goal of integrating environmental factors into the broader systems of racial capitalism that have taken shape with the rise of neoliberal Asian manufacturing and labor diasporas. By developing a method for analyzing how oil-driven logics of transnationalism have produced both population flows and environmental harms, it is my hope that this book has also demonstrated that it is possible to offer a systemic account of how such equalities emerge internationally without separating environmental from economic, social, and political forces. While the particular, the local, and even the individual contexts of migration and environmental harm must be explored in accounts of contemporary climate change, moving from particular to planetary contexts—and in the process developing geographical knowledge about how local problems are scaled up in environmental discourse—is one of the most difficult tasks of critical social methods that analyze migration, security, and environmental risk. Unlike the security thinking arising from border imperialism, which tends to reinforce crisis narratives and the stereotyped discourses of difference reifying a nationalist or statist worldview, thinking about how the carbon economy, racial capitalism, and neoliberalism intersect in developing contemporary forms of international inequality and violence is vital when developing transnational methods for connecting shared struggles of people affected by agricultural collapse and weather disaster.

Such an approach can help us work through some of the challenges that have emerged in scholarship that attempts to integrate environmental concerns with critical social theories. Critical refugee studies, for one, reflects a growing interest in how environmental questions intersect with social ones. Since the field is invested in rethinking the normative legal categories through which migration is publicly narrated and interpreted, in the process interrogating the presumption that receiving states confer a gift of freedom on the refugee, thinking about environmental forces in migration may help the field articulate the complexity of refugee lifeworlds and the political potential of refugee critique. By refusing the legal distinctions separating migrant from refugee and political from nonpolitical displacement forces, critical refugee studies in turn makes room for thinking about the long-term, differently unfolding crises that affect access to water, food, employment, and other resources in climate-affected regions. However, it will be necessary for such methods to follow the complexity of social and environmental factors and insist on a historical approach to borders, race, and labor in order to understand the imbrication of environmental processes in longer political economies of displacement. At the same time, breaking down "environment" or "climate" into distinct ecological processes such as drought and coastal salinization can help demonstrate how specific zones of ecological change are recruited into national and transnational migration pathways. Rather than assimilating "climate refugees" as an iconic new alternative to traditional definitions of the refugee, the field can insist on a community-centered cartography of displacement attentive to the ways in which environmental processes enter into practices of social reproduction in agrarian zones and other communities rendered vulnerable to neoliberal circuits of debt and displacement.

At the same time, the figure of the climate migrant presses Asian American studies to grapple with the complex ways in which transnational migration flows structure forms of U.S. imperial securitization and the formations of racial capitalism that emerge with changing relations between North America, the Gulf states, the Pacific, and East Asia. If the geographic shifts in international finance have generated visions of a New Asian Century, the figure of the climate refugee as an underside of economic globalization suggests that the field's attention to growing global inequalities and new forms of militarism and colonization in the Pacific can be generative for reflecting on how environmental phenomena is ambivalently positioned in such transcontinental visions of global interconnection at a moment in which U.S. empire has been challenged on a variety of fronts.

The roles of both critical refugee studies and Asian American studies in expanding such a transnational perspective is especially critical in this moment of rising fascism worldwide, intensified by the inequalities of the COVID-19 pandemic. On the one hand, the interlinked health, environmental, economic, and political crises of the moment require activists and scholars to develop an understanding of how the rise of fascist leaders emerges from the combination of post-9/11 militarized security and longer-term neoliberal transitions that shift the economic bases of nationalism. The retrenchment of provincial and national borders under emergency public health authority has created complex pressures for migrants and their natal communities. If in India we have witnessed the massive expulsion of millions of migrant workers from the cities, in other locations enhanced border enforcement and visa restrictions are forcing increasingly perilous journeys for workers and other displaced peoples.

On the other hand, it is not enough to narrate how this neoliberal circuit of transnationalization and nationalization generates persistent crisis. We cannot understand the racialized violence of the carbon economy without emphasizing the interspecies and infrastructural complexity of interrelated environmental disasters such as climate change and the COVID-19 pandemic. As discussed in chapter 2, between the SARS outbreaks of 2002–4 and the COVID-19 outbreak of December 2019 to the present, China experienced what may be the largest and most rapid urbanization process in history.[3] The number of urban residents in the country grew to over 670 million, as manufacturing and transit networks spread across the country, including to central China, where the COVID-19 outbreak began. At the crossroads of many migrants who shuttle between the cities and agrarian natal communities, Wuhan had become a transit hub, a center of rapid development, and a growing industrial region. Even as such development encroaches on habitats for bats, which provide a natural reservoir for coronaviruses, growing migration transit pathways help to conduct possible zoonoses farther afield on the routes of the supply chain. To the extent that the pandemic was generated by growth in migration and transnational travel, it emerged from the same energy and environmental relations that I have discussed throughout this book.[4] The interspecies basis of the current pandemic travels along routes of environmental destruction that intensify pressures on agrarian zones and render them increasingly vulnerable to forms of environmental surveillance that disavow the role that consumption in urban centers plays in spaces configured as potential sites of zoonotic propagation.[5]

In order to explore how political and economic transformations are linked to environmental ones through the centering of racial capitalism in Asian trade, finance, and labor circuits, it is necessary to engage with some geographic questions that have been historically taken up by Asian studies scholars, at times critically reflecting on the limits of Cold War influences on such area studies. However, this book has also asserted that the reproduction of the capitalist system through its transnationalization into Asian migration and extractive circuits has had more widely distributed impacts that require Asian studies and Asian American studies to rethink the cartographic outlines of the fields. The rise of climate security discourses and the circulation of the climate refugee as both a legal category and a media icon have occurred alongside rapidly growing carbon emissions that affect locations across the globe. By stressing the key role of labor, oil, and production in Asia to these transborder forces, I do not mean to sideline work that attends to other regional dimensions of racial capitalism or environmental crisis. Given the regularization of weather disasters over the past two decades, there is likely to be much more future scholarship exploring the specific ecological effects of climate change as they intersect with varied crises of social reproduction.

Although these effects are worldwide, they also map onto long-standing settler colonial violences of racial capitalism. The debates among comparative racialization scholars that stress the foundational violences of settler colonialism and anti-Black labor regimes emphasize the key role that Atlantic processes of commoditized land and labor played in the making of the modern system of racial capitalism. Comparative racialization studies—which work to disentangle the ways that forces like warfare, colonization, labor exploitation, and debt forge particular regimes of racial power that differentially affect Indigenous and racially minoritized groups—have much to contribute to tracing the situated differences and divergences in climate change impacts. This is particularly important today, since finance weaves these other forces of racial violence into a speculative economy that reinforces indebtedness across varied contexts of accumulation. It will be a key challenge for comparative racialization studies in an era of climate change to confront how diminishing land bases and growing surplus labor are revalued through circuits of finance and massified agrarian debt in the South, as well as through climate adaptation schemes that tend to favor migration and urbanization as solutions to poverty.

At the same time, the differences in context that produce some of the key debates in comparative racialization studies—for example, the differences in the ways that Indigenous land and Black labor have been historically appropriated by colonial racial regimes—can at times devolve into a kind of exceptionalism that moves away from systemic analysis of the manner in which war, land, labor, and finance interrelate in racial capitalism. The racial hierarchies of labor discussed in this book have at times been modeled on anti-Black strategies of labor exploitation, as discussed in the histories of Aramco oil production and the Gulf kafala labor system. But it is important to not simply view Asian labor diasporas as mimetic copies of an original exclusionary labor regime aimed at other groups.[6] Nor is it enough to note the relative privilege of some Asian immigrants in settler colonial contexts, wherein Asian settler colonialism emerges as a secondary violence following white settlement.[7] It is possible to critique complicity with structures of settler colonial racialization within Asian diasporic communities without reifying xenophobic stereotypes of Asian migration as a sign of the growing domination of Asian nation-states in the world system. Taking account of how neo-fascist depictions of the "rise of Asia" have made "Asia" itself an overburdened signifier in many contexts of political debate is critical as comparative racialization scholars and Asian diasporic activists reflect on emerging circuits of migration. Comparative racialization studies is best poised to develop an integrated account of racial capitalism in the present if it subjects exceptionalist claims about the durability of racial structure to critical analysis by tracking how economic transformations bring new racial regimes and meanings into relation.[8] This book has thus stressed the interdependencies of Asian production, finance, and labor with neoliberal, carbon-fueled circuits of U.S. settler colonialism and its resulting border imperialism. Such a line of thinking helps us understand the persistence of racialized disposability in this transitional moment where U.S. power appears threatened even as some aspects of U.S. domination are structured into the reproductive forces of racial capitalism.

Critical migration studies and critical race theories are in turn necessary for developing methods to address the social violences of planetary forms of environmental change. This is especially so as they address the ways that race configures the very divide between society and nature—the ethnoclass struggles that Sylvia Wynter defined as the basis of our public debates over the environment. They furthermore help environmental racism scholars address the ways that ideas about managing waste also configure emerging turns to environmental governance that, this book argues, reinforce international inequalities. Although northern spaces of industry and urbanization—as sites of development of ecologically razed settlements, extractive technologies, and massified forms of capitalist circulation—might themselves be seen as the primary zones of environmental degradation, the contradictory racialization of southern environments as both carbon sinks and wastelands—sites that must be alternately conserved and developed—maintains a focus on climate-affected agrarian environments as sites of likely intervention by the forces of neocolonial development or conservation. As such, critical refugee studies and critical border studies can benefit from thinking about environmental racisms that subject migrants to transborder forms of ecological harm, even as such forms of harm are often determined by basic inequalities of infrastructure, access to land and property, and vulnerability to toxicity and other negative effects of capitalist waste.

Studies at the intersection of critical race studies and critical security studies are, finally, another fruitful intersection for understanding how race configures strategies of governance in the present context of migration. Groups configured as security risks or mobilized as Indigenous adaptation models are likely to face compounding forces of displacement generated not only by extractive enterprise but also by the speculative governance of climate adaptation schemes. The connection between efforts to account for climate migration as an environmental injustice and the potential to mobilize race within new governance techniques of human security is one potential outcome of the collision of security and migrant rights efforts. The manner in which this may proliferate calculation of environmental harms as a speculative race-making technology may still play out in a variety of directions. Although such accounting of differential vulnerability to disaster could be used to redistribute state resources to particular groups facing ecological change, it also portends new forms of surveillance and disaster intervention that have unequal effects based on factors like citizenship/documentation status and conditional resettlement funding that requires populations to move away from coastal or agrarian home regions. Environmental racism research is necessary to track how displacement is used for such race-making projects and to consider how this configures migration as a sign of vulnerability, an adaptation strategy, a node of human capital, or a condition of surveillance in different contexts.

Overall, environmental knowledge must be configured as part of a larger picture of racial capitalism attending to the historical trajectories of migration, the systemic causes of environmental destruction, and the speculative horizons of climate finance and security. Emphasizing climate change's effects on social systems without analyzing the feedback loops and interrelations with political economy may offer a strategic shortcut to challenging climate denialism among corporate elites and right-wing supporters of continued fossil fuel use. However, to overemphasize the potential of environmental thinking by itself to undo humanist conceits at the heart of liberalism can run into complicated roadblocks, especially when invoking "human" environmental agencies without attending to the complexity of how the human is conceived and crosscut by inequality, or even how such differences are increasingly mobilized in human security practices.[9] To the extent that climate adaptation schemes engage with crisis thinking and techniques of securitization, they reflect how climate change is itself a site of narrative worldmaking, a proliferation of speculative visions of human extinction that anticipate life itself as a shrinking zone of habitation.

In contrast to the securitizing forces of racial capitalism and border imperialism, many of the social movement responses to weather disaster involve mutual aid and the attempt by climate-affected peoples to make connections to struggles elsewhere—for example, in the agricultural cooperatives of Rojava or the attempts to model mutual aid among hurricane-affected coastal peoples from Puerto Rico to Bangladesh and points in between. If collective responses to climate change are able to produce fundamental resources to challenge current configurations of racial capitalism, it is likely that they will do so through the development of networks of solidarity that both deliver material resources beyond the purview of state visions of aid (attached, as they are, to inequalities of citizenship and access) and build on decades of resistance to neoliberal transnationalism. The social movements that in the early 2000s challenged the international debt regime and structural adjustment requirements produced intersections between organizations focused on Indigenous agrarian struggles, hunger, environmental violence, animal liberation, and labor conditions, articulating models for later convergences, such as the World Social Forum, the Arab uprisings, and Occupy mobilizations. To the extent that such developments have been accompanied by a revanchist formation of rising fascism in many countries should come as no surprise, as the very condition of neoliberal accumulation has been the massive expansion of wealth inequalities among and between states, a situation that energizes nostalgic invocations of nationalism, sexism, and racism in varied political contexts.

This is especially important given my emphasis in chapter 1 on the emerging right-wing interest in climate migration, including a new political opening for green nationalism. Although political parties and formations on the Right have, since the 1990s, tended to oppose stronger environmental regulation and undermined public understanding of climate science, the Far-Right formations—including the Alt-Right and white supremacist groups—are increasingly finding that xenophobia is more successful politically than anti-environmentalism. Climate migration discourse forms the key bridge between environmental thought and xenophobia that may allow a shift in the manner in which the Right addresses climate change. For anyone who has observed the longer histories in which the nostalgic romanticization of national environments—from the United States to Germany to India, as only a few examples—has been yoked to notions of racial identity and national superiority, it should come as no surprise that environmentalism can be entirely compatible with fascism. Today, dozens of countries have erected border walls to keep out migrants, and the idea that environmental disasters might stoke racial animus and social crisis is one of the primary ways that the Right may be able to capitalize on disasters to advance its growing influence over state policies and media discourse. If in the early 1990s Etienne Balibar was correct in suggesting that postcolonial migration to Europe reflected a "neo-racism" that reframed racism as an inevitable outcome of increasing transborder contact and ethnic assimilation, climate migration discourse offers a potential discursive site for embedding the normalization of racism and xenophobia, configured as social realist responses to a world of massifying population and socio-ecological complexity.[10] Rather than doubling down on the turn to security and to nationalism in responses by the Left to such neo-racist discourse, it is more useful for anti-racist, environmental, and immigrant movements to forcefully critique and challenge security thinking.

Given this situation, *Planetary Specters* argues that oil-fueled debt and environmental change helps to regenerate some contours of race and racism, primarily through the shifting geographies of rural-urban mobility and remittance and through the acceleration of major inequalities in the world system. Neoliberalism is not an independent economic logic but a reproductive system reliant on racialized management of labor, disposing of some workers through intricate physical mobilities and telemobilities while recruiting other segments of vulnerable labor and extractable land. As long as oil supports a deeply unequal international financial structure, combined with accelerating environmental catastrophe, it will continue to exacerbate forms of human mobility that reproduce the unequal vulnerability of racialized agrarian peripheries. To the extent that these peripheries have been collected into forms of governance and securitization that blur migration with war, they have been subjected to neocolonial adaptation interventions emphasizing debility, underdevelopment, and the potential for conflict. These associations in turn may be mobilized to ensure that speculated displacements take place in advance, that vulnerability be rescripted as adaptive capacity by simply declaring the unlivability of the extractive zone. As such, racism helps to devalue Indigenous and agrarian spaces that are written off as less productive than spaces of production. Such developments may extend into new forms of green valuation, which for the time being remain vested in the high price attributed to fossil fuels and the interrelation between carbon consumption and the development of human capital.

To the extent that focus on the figure of the climate refugee appears new, it is also embedded in the distinct configurations of neoliberal difference that situate race itself in the structures of communication and media. Tracking this development benefits from analyzing race in relation to gender, sex, and disability using approaches to understanding rights and humanitarianism from feminist, queer, and disability studies. The icon of the debilitated climate migrant—from Ioane Teitiota in chapter 1 to Alan Kurdi in chapter 4—thus teaches us something about how the vulnerable body provides affective scaffolding for the racialization of migrants today. Such is the case in the framing of climate migration as a scene of debilitation threatening security, as well as in more hopeful visions of climate adaptation as overcoming such debility. Icons of vulnerability or activist transcendence may help to distill activist struggles around a key figure signaling transformation of a social order, but they also accrue symbolic meanings that may be turned against various forms of liberatory struggle.[11] If social media was once championed as a tool for social change by some activists critiquing mainstream media discourses, the deterritorialization of digital media in the present moment configures ever more detailed population constructions and iconic associations that may be mobilized for control. This is not an intractable problem but one that energizes spurious discourses of climate migration, the form of which is at times overdetermined by the platform. Entrenched in the liberal mediasphere, climate migration stories shared on platforms like Twitter and Facebook transform migrant bodies into digital avatars of blowback produced by the fossil-fueled corporate nationalisms they seek to transcend via figures of humanitarian rescue. As refugee stories and migrant crisis discourses occupy an increasing online presence in traditional and social media, more research will need to examine how digital infrastructures affect securitization, migration, and environmental processes alike.

This is one reason why attempts to intervene in climate migration discourses like Mejía's digital *testimonio* are necessary for developing integrated analysis of racial capitalism that will address systemic factors of displacement. Such efforts are only one step in combating the lengthy history of neocolonial representations of southern peoples and environments, but they show us how the intertwined geographies of communication, environmental change, and economic structure are central to the political challenges raised by the specters of climate change, economic inequality, and rising fascism in the twenty-first century.

Acknowledgments

This book took its final form as I began teaching, writing, and advocating alongside the students, staff, and faculty in the Critical Race and Ethnic Studies (CRES) program at the University of California, Santa Cruz. Thanks to Dana Ahern, Taylor Ainslie, Chrissy Anderson-Zavala, Neda Atanasoski, Courtney Bonam, Micha Cárdenas, Vilashini Cooppan, A. M. Darke, Camilla Hawthorne, Christine Hong, Talib Jabbar, Jenny Kelley, Jane Komori, Shauntay Larkins, Adonay Lozano Moreno, Nick Mitchell, Nidhi Mahajan, Trung Nguyen, Marcia Ochoa, Emily Padilla, Aitanna Parker, Juan Poblete, Eric Porter, Felicity Amaya Schaeffer, Jared Semana, Savannah Shange, Anney Treymanny, Ronaldo Wilson, Karen Tei Yamashita, Alice Yang, Ka-Eul Yoo, and Jerry Zee for their commitment to promoting CRES, building its community, and insisting on its orientation to transform the institution despite all odds. Special thanks to Neda, Christine, Nick, and Taylor for their leadership during the most difficult moments of administrative foot-dragging, and to countless graduate student strikers and members of the People's Coalition who worked to collectively envision a new relationship to the university amidst the wildcat strike of 2019.

Thanks to numerous comrades in North Carolina, whose organizing against racist monuments, policing, and immigration surveillance has provided inspiration for this book. I thank a number of colleagues at other universities, including Aimee Bahng, Mabel Gergan, Jairus Grove, Dana Luciano, Michael Lundblad, Renisa Mawani, Jasbir Puar, Malini Ranganathan, Trisha Remitir, Pavithra Vasudevan, and Priscilla Wald, whose discussions of environment, species, and colonialism have been generative for this study. I benefited from feedback I received from audiences at Dartmouth College, the University of Colorado–Boulder, Wesleyan University, American University, and Georgetown University. Different versions of sections of the manuscript have been published or are forthcoming as "Race, Human Security, and the Climate Refugee," *English Language Notes* 54, no. 2 (2016): 25–32; "Figuring the Climate Refugee: From Insecurity to Adaptation in Representations of Bangladeshi Environmental Migration," in *Insecurity*, ed. Richard Grusin and Maureen Ryan (Minneapolis: University of Minnesota Press, forthcoming); and "Weather as War: Race, Disability, and Environmental Determinism in the Syrian Climate Wars Thesis," *Critical Ethnic Studies* 6:1 (2020): https://doi.org/10.5749/CES.0601.ahuja. I was extremely lucky to have the guidance of Dylan White at the University of North Carolina Press, who oversaw the editorial process and carefully attended to all the details. I give special thanks to Renisa Mawani and Priscilla Wald, whose careful reviews of the manuscript helped me clarify the arguments of the book in its final stages.

Finally, I have benefited from the direct support of many close friends and family members during the writing of this book. Special thanks to Hong-An, Dwayne, Xuân June, Jes, Sarah, Nina, Antony, Eliza, Jenny, Dylan, Trisha, Chelsea, Jess, Pavithra,

Atiya, Armond, and Felicity for sustaining community over these past few years. Thanks also to my family, including Sain, Usha, Liliana, Radoslav, Sonia, Shyla, Sarita, and Barney—and to Neda, Naya, and Persia for filling each day with surprise, laughter, and love. This book is dedicated to Naya in the hope that you may one day inhabit a more just world, one in which the wonders of life around us no longer sacrificed in hollow pursuits of profit and domination.

Notes

Introduction

1. Wikipedia, s.v. "Central American migrant caravans: Late 2018 caravans," last modified January 17, 2020, 17:48, https://en.wikipedia.org/wiki/Central_American_migrant_caravans#Late_2018_caravans.

2. Oliver Milman, Emily Holden, and David Agren, "The Unseen Driver behind the Migrant Caravan: Climate Change," *Guardian*, October 30, 2018, www.theguardian.com/world/2018/oct/30/migrant-caravan-causes-climate-change-central-america.

3. Milman, Holden, and Agren, "Unseen Driver behind the Migrant Caravan."

4. Jason Cons, "Global Flooding," *Anthropology Now* 9, no. 3 (2017): 47–52.

5. Cedric Robinson, *Black Marxism: The Making of the Black Radical Tradition* (Chapel Hill: University of North Carolina Press, 1983), 2.

6. Aimee Bahng, *Migrant Futures: Decolonizing Speculation in Financial Times* (Durham, NC: Duke University Press, 2018), 119–45.

7. Nicholas De Genova, "The 'Migrant Crisis' as Racial Crisis: Do *Black Lives Matter* in Europe?" *Ethnic and Racial Studies* 41, no. 10 (2018): 1765–82.

8. Harsha Walia, *Undoing Border Imperialism* (Oakland: AK Press/Institute for Anarchist Studies, 2013), 11.

9. Suyapa Portillo Villeda and Gerardo Torres Zelaya, "Why Are Honduran Children Leaving?," *CounterPunch*, June 27, 2014, www.counterpunch.org/2014/06/27/why-are-honduran-children-leaving.

10. Yermi Brenner and Katrin Ohlendorf, "Time for the Facts: What Do We Know about Cologne Four Months Later?" *De Correspondent*, May 2, 2016, https://thecorrespondent.com/4401/time-for-the-facts-what-do-we-know-about-cologne-four-months-later/1073698080444-e20ada1b.

11. Brian Massumi, "National Enterprise Emergency: Steps toward an Ecology of Powers," *Theory Culture and Society* 26, no. 6 (2009): 153–85; Gregoire Chamayou, *A Theory of the Drone*, trans. Janet Lloyd (New York: New Press, 2015).

12. Neda Atanasoski, *Humanitarian Violence: The US Deployment of Diversity* (Minneapolis: University of Minnesota Press, 2013).

13. Junaid Rana, "The Story of Islamophobia," *Souls* 9, no. 2 (2007): 148–61; Atiya Husain, "Retrieving the Religion in Racialization: A Critical Review," *Sociology Compass* 11, no. 3 (2017): e12507, https://doi.org/10.1111/soc4.12507.

14. "The Biden Plan for a Clean Energy Revolution and Environmental Justice," accessed December 1, 2020, https://joebiden.com/climate-plan/#. Thanks to Aimee Bahng for sharing this citation.

15. International Energy Association, "Global Energy Review 2020," April 2020, www.iea.org/reports/global-energy-review-2020. See also Michael T. Klare, "Covid-19

Is Forcing Us to Rethink How We Consume Energy," *Nation*, April 29, 2020, www.thenation.com/article/environment/coronavirus-oil-energy-consumption.

16. U.S. Energy Information Administration, "EIA Projects 28% Increase in World Energy Use by 2040," September 14, 2017, www.eia.gov/todayinenergy/detail.php?id=32912.

17. Philip Alston, *Climate Change and Poverty: Report of the Special Rapporteur on Extreme Poverty and Human Rights*, UN Human Rights Council, 41st sess., June 24, 2019, https://undocs.org/pdf?symbol=en/A/HRC/41/39.

18. The World Bank, for example, forecasts that the pandemic will only temporarily cause migration flows to drop. See *COVID-19 Crisis through a Migration Lens*, Migration and Development Brief 32 (Washington: World Bank Group, 2020), https://openknowledge.worldbank.org/bitstream/handle/10986/33634/COVID-19-Crisis-Through-a-Migration-Lens.pdf?sequence=5&isAllowed=y.

19. Cons, "Global Flooding," 48.

20. Etienne Piguet, "From 'Primitive Migration' to 'Climate Refugees': The Curious Fate of the Natural Environment in Migration Studies," *Annals of the Association of American Geographers* 103, no. 14 (2013): 148–62.

21. Shweta Jayawardhan, "Vulnerability and Climate Change Induced Human Displacement," *Consilience* 17 (2017): 103–42.

22. See Liisa Malkki, "Speechless Emissaries: Refugees, Humanitarianism, and Dehistoricization," *Cultural Anthropology* 11, no. 3 (1996): 377–404.

23. Ruth Wilson Gilmore, *Golden Gulag: Prisons, Surplus, Opposition, and Crisis in Globalizing California* (Berkeley: University of California Press, 2007), 28.

24. Walia, *Undoing Border Imperialism*, 34–35.

25. Randall Williams, *The Divided World: Human Rights and Its Violence* (Minneapolis: University of Minnesota Press, 2010).

26. See Patrick Wolfe, *Traces of History: Elementary Structures of Race* (London: Verso, 2016). Wolfe's description of racial regimes as central avenues through which settler colonial states manage demographics of labor and land is useful for comparative analysis of how race and capitalism intertwine to target minoritized groups. However, Wolfe's reliance on assessment of human reproduction as a metric for the value of labor requires further analysis, as it tends to gloss over the sexual violence inherent in systems of slavery and indenture. For a critique of the centrality of Black women's reproductive labor to the U.S. racial regime, see Hortense Spillers, "Mama's Baby, Papa's Maybe: An American Grammar Book," *Diacritics* 17, no. 2 (1987): 64–81. Furthermore, the schematic division between Wolfe's concept of elimination (used to accumulate native land) and exclusion (used to accumulate racialized labor) can preclude nuanced contextual analysis of multitiered racial hierarchies and phenomena, such as the exploitation of native labor. For an alternative approach that integrates land and labor questions, see Tomás Almaguer, *Racial Fault Lines: The Historical Origins of White Supremacy in California* (Berkeley: University of California Press, 1994); for an important if polemical attempt to synthesize multiple colonial regimes of race and labor into a reading of U.S.-Mexico border politics, see Rosaura Sánchez and Beatrice Pita, "Rethinking Settler Colonialism," *American Quarterly* 66, no. 4 (2014): 1039–55.

27. Macarena Gomez-Barris, *The Extractive Zone: Social Ecologies and Decolonial Perspectives* (Durham, NC: Duke University Press, 2007), 2.

28. Jason Moore, *Capitalism in the Web of Life: Ecology and the Accumulation of Capital* (London: Verso, 2015).

29. On the demography of European genocide in the Pacific and the Americas, see David Stannard, *American Holocaust: The Conquest of the New World* (Oxford: Oxford University Press, 1992). On the environmental history of early colonial settlement in the Americas, see Alfred Crosby, *The Columbian Exchange: Biological and Cultural Consequences of 1492* (Westport, CT: Greenwood, 1972). On Indigenous critiques of anthropocentrism as a settler colonial ideology, see Jodi Byrd, *The Transit of Empire: Indigenous Critiques of Colonialism* (Minneapolis: University of Minnesota Press, 2011); Kyle Whyte, "Our Ancestors' Dystopia Now: Indigenous Conservation and the Anthropocene," *Routledge Companion to Environmental Studies*, ed. U. Heise, J. Christensen, and M. Niemann (New York: Routledge, 2017), 206–15; Nick Estes and Jaskiran Dhillon, eds., *Standing with Standing Rock: Voices from the #NoDAPL Movement* (Minneapolis: University of Minnesota Press, 2019); Glen Coulthard, *Red Skins, White Masks: Rejecting the Colonial Politics of Recognition* (Minneapolis: University of Minnesota Press, 2014).

30. Neel Ahuja, "The Anthropocene Debate: On the Limits of Colonial Geology," September 9, 2016, https://ahuja.sites.ucsc.edu/2016/09/09/the-anthropocene-debate-on-the-limits-of-colonial-geology.

31. Lisa Lowe, *The Intimacies of Four Continents* (Durham, NC: Duke University Press, 2015).

32. Denise Ferreira da Silva, *Toward a Global Idea of Race* (Minneapolis: University of Minnesota Press, 2007); Kathryn Yusoff, *A Billion Black Anthropocenes* (Minneapolis: University of Minnesota Press, 2018).

33. Amitav Ghosh, *The Great Derangement: Climate Change and the Unthinkable* (Chicago: University of Chicago Press, 2017).

34. Andreas Malm, *Fossil Capital: The Rise of Steam Power and the Roots of Global Warming* (London: Verso, 2015).

35. Timothy Mitchell, *Carbon Democracy: Political Power in the Age of Oil* (London: Verso, 2011).

36. Vijay Prashad, *The Darker Nations: A People's History of the Third World* (New York: New Press, 2008).

37. Paul Crutzen and Eugene Stoermer, "The 'Anthropocene,'" *Global Change Newsletter* 41 (2000): 17; Dipesh Chakrabarty, "The Climate of History: Four Theses," *Critical Inquiry* 35 (2009): 197–222.

38. For examples of widely cited theoretical texts, see Chakrabarty, "The Climate of History; Timothy Morton, *Hyperobjects: Philosophy and Ecology after the End of the World* (Minneapolis: University of Minnesota Press, 2013); Malm, *Fossil Capital*.

39. Yusoff, *Billion Black*; Ghosh, *Great Derangement*; Whyte, "Our Ancestors' Dystopia Now," 206–15. I have explored such critiques in more detail in Ahuja, "Anthropocene Debate"; Neel Ahuja, "Intimate Atmospheres: Queer Theory in a Time of Extinctions," *GLQ* 21, no. 2–3 (2015): 365–85; Neel Ahuja, "Posthuman New York: Ground Zero of

the Anthropocene," *Animalities: Literary and Cultural Studies Beyond the Human*, ed. Michael Lundblad (Edinburgh: Edinburgh University Press, 2017), 237–52.

40. Jason Moore, *Capitalism in the Web of Life: Ecology and the Accumulation of Capital* (London: Verso, 2015); Jason Moore, ed., *Anthropocene or Capitalocene? Nature, History, and the Crisis of Capitalism* (Oakland: PM Press, 2016); Donna Haraway, Anna Tsing, and Gregg Mitman, *Reflections on the Plantationocene*, June 18, 2019, https://edgeeffects.net/wp-content/uploads/2019/06/PlantationoceneReflections_Haraway_Tsing.pdf; Jairus Grove, "Response: The New Nature," *Boston Review*, January 11, 2016, http://bostonreview.net/forum/new-nature/jairus-grove-jairus-grove-response-jedediah-purdy.

41. Alexander Weheliye, *Habeas Viscus: Racializing Assemblages, Biopolitics, and Black Feminist Theories of the Human* (Durham, NC: Duke University Press, 2014); Zakiyyah Iman Jackson, *Becoming Human: Matter and Meaning in an Antiblack World* (New York: New York University Press, 2020); Neel Ahuja, *Bioinsecurities: Disease Interventions, Empire, and the Government of Species* (Durham: Duke University Press, 2016), esp. xiii–xv.

42. On the layering of these two affective realities of expansion and contraction within lived geographies of neoliberal "globalization," see Ruth Wilson Gilmore, "Race and Globalization," in *Geographies of Global Change: Remapping the World*, ed. R. J. Johnston, Peter Taylor, and Michael Watts, 2nd ed. (Malden: Blackwell, 2002), 261–74.

43. For an important critique of the "great acceleration," see Elizabeth M. DeLoughrey, *Allegories of the Anthropocene* (Durham, NC: Duke University Press, 2019), esp. 98–132.

44. Duncan Green, *Silent Revolution: The Rise and Crisis of Market Economics in Latin America* (New York: Monthly Review, 1995); Samir Amin, *Capitalism in the Age of Globalization: The Management of Contemporary Society* (London: Zed, 1997).

45. Akhil Gupta, *Postcolonial Developments* (Durham, NC: Duke University Press, 1998).

46. Michael Watt, "Resource Curse? Governmentality, Oil, and Power in the Niger Delta, Nigeria," *Geopolitics* 9, no. 1 (2004): 50–80.

47. Richard Nixon, "Special Message to the Congress on Energy Resources," quoted in Timothy Mitchell, *Carbon Democracy: Political Power in the Age of Oil* (London: Verso, 2011), 191.

48. Mitchell, *Carbon Democracy*, 250.

49. Mitchell, *Carbon Democracy*, 250.

50. Imre Szeman, "How to Know about Oil: Energy Epistemologies and Political Futures," *Journal of Canadian Studies* 47, no. 3 (2013): 145–68 (quote on 146).

51. Stephanie LeMenager, *Living Oil: Petroleum Culture in the American Century* (Oxford: Oxford University Press, 2014); Edward Burtynsky, *Oil* (Steidl, 2009); Amitav Ghosh, "Petrofiction: The Oil Encounter and the Novel," *New Republic*, March 2, 1992, 29–33. For an important critical review of these works, see Brent Ryan Bellamy, "The Aesthetic Textuality of Oil," in *The Palgrave Handbook of Twentieth and Twenty-First Century Literature and Science*, ed. Triangle Collective (Cham, Switzerland: Palgrave Macmillan, 2020), 63–77.

52. Christopher Sall and Urvashi Narain, "Air Pollution: Impact on Human Health and Wealth," *The Changing Wealth of Nations 2018: Building a Sustainable Future* (Washington, DC: World Bank Group, 2018), 171–88.

53. Kanta Kumari Rigaud, Alex de Sherbinin, Bryan Jones, Jonas Bergmann, Viviane Clement, Kayly Ober, Jacob Schewe, et al., *Groundswell: Preparing for Internal Climate Migration* (Washington, DC: World Bank, 2018), 25–26.

54. Reza Negarestani, *Cyclonopedia: Complicity with Anonymous Materials* (Melbourne: re.press, 2008), 27.

55. Christian Parenti, *Tropic of Chaos: Climate Change and the New Geography of Violence* (New York: Nation Books, 2012).

56. Randall Williams, *The Divided World: Human Rights and Its Violence* (Minneapolis: University of Minnesota Press, 2010).

57. Sylvia Wynter, "Unsettling the Coloniality of Being/Power/Truth/Freedom: Towards the Human, After Man, Its Overrepresentation—an Argument," *CR: The New Centennial Review* 3, no. 3 (2003): 260–61.

58. Both oil and refugees became a central preoccupation in Foucault's writings in the late 1970s and early 1980s as he responded to the Iranian Revolution of 1979 and the aftermath of the Vietnam War. Disappointed that the North Vietnamese struggle against U.S. and French empires had generated the spectacle of the "boat people"—the hundreds of thousands of refugees who fled by boat in the wake of the war—Foucault sought other models of revolution elsewhere in the Global South. Foucault claimed that the oil strikes at Abadan, Iran, reflected a new "political spirituality," a religiosity that articulated what could become a new, non-Western model of revolutionary action. See Michel Foucault, "What Are the Iranians Dreaming About?" in *Foucault and the Iranian Revolution: Gender and the Seductions of Islamism*, ed. Janet Afary and Kevin Anderson (Chicago: University of Chicago Press, 2014), 203–8. For Foucault, the Islamist contingent of the Iranian revolutionaries "struggle to present a different way of thinking about social and political organization, one that takes nothing from Western philosophy, from its juridical and revolutionary foundations. In other words, they try to present an alternative based on Islamic teachings." Foucault, "Dialogue between Michel Foucault and Baqir Parham," in Afary and Anderson, *Foucault and the Iranian Revolution*, 185–86. Representing oil strikers as part of a unified and, for him, refreshingly nonsecular force against U.S. hegemony, Foucault reports that they took on Abadan, the "biggest refinery in the world," a site no "European has not dreamed about." Foucault, "The Revolt in Iran Spreads on Cassette Tapes," in Afary and Anderson, *Foucault and the Iranian Revolution*, 217.

Such romantic invocations of Islamic revolution would not hold water for long. Foucault's idealization of the Islamic character of political action came under pressure as the revolution progressed—particularly with the emergence of enforced veiling and repression of dissent by the newly founded oil-rich Islamic republic. Foucault turned away from this model of political spirituality toward an embrace of certain discourses of humanism, internationalism, and even interventionism. Foucault's eventual argument for protection against torture in his Iran writings anticipates a more open embrace of certain human rights rhetorics emergent with international NGOs like Amnesty International and Doctors without Borders, an organization founded by Foucault's activist collaborator and former French foreign secretary Bernard Kouchner. Jessica Whyte points to a more elaborated claim for rights—and a neocolonial right to intervene—in Foucault's speech in defense of the so-called boat people at the UN in 1981. In it he invokes "an international citizenship that has its rights and its duties," which include the duties of individuals to "wrench from governments the monopolization of the power to effectively intervene." As Whyte argues, Foucault's articulation of the right to intervene was explicitly invoked by France in later international contexts that began to shape a neocolonial right of humanitarian intervention. Jessica Whyte, "Foucault's 'Distrust of Legalism': On Human Rights and the Revolution in Iran," in *Law and Philosophical Theory*, ed. Thanos Zardalouis (Lanham, MD: Rowman and Littlefield, 2018), 27–44; Jessica Whyte, "Human Rights: Confronting Governments? Michel Foucault and the Right to Intervene," in *New Critical Legal Thinking: Law and the Political*, ed. Matthew Stone, Illan rua Wall, and Costas Douzinas (New York: Routledge, 2012), 11–31.

59. Craig Welch, "Climate Change Helped Spark Syrian War, Study Says," *National Geographic*, March 2, 2015, www.nationalgeographic.com/news/2015/3/150302-syria-war-climate-change-drought.

Chapter One

1. Kanta Kumari Ringaud, Alex de Sherbinin, Bryan Jones, Jonas Bergmann, Viviane Clement, Kayly Ober, Jacob Schewe, et al., *Groundswell: Preparing for Internal Climate Migration* (Washington, DC: World Bank, 2018), 1.

2. J. R. McNeill and Peter Engelke, *The Great Acceleration: An Environmental History of the Anthropocene since 1945* (Cambridge, MA: Harvard University Press, 2016).

3. See Lisa Lowe, *The Intimacies of Four Continents* (Durham, NC: Duke University Press, 2015).

4. "2019 Internal Displacement Figures by Country," Global Internal Displacement Database, Internal Displacement Monitoring Centre, accessed July 9, 2019, www.internal-displacement.org/database/displacement-data. I added IDMC's data for the top five countries in terms of disaster-displaced population for 2018. This data reports 12,429,000 internally displaced people in Philippines, China, India, United States, and Indonesia.

5. Essam el-Hinnawi, *Environmental Refugees* (Nairobi: UNEP, 1985), 22.

6. Myers's series of essays on the topic follows this trajectory of discussion of climate change as first a humanitarian disaster and then a security risk. See Norman Myers, "Environmental Refugees in a Globally Warmed World," *BioScience* 43, no. 11 (1993): 752–61; Norman Myers, "Environmental Refugees: An Emergent Security Issue," 13th Economic Forum, Prague, May 23–27, 2005.

7. Casey Williams, "What Happens When the Alt-Right Believes in Climate Change?" *Jewish Currents*, August 13, 2018, https://jewishcurrents.org/what-happens-when-alt-right-believes-climate-change.

8. Kate Aronoff, "The European Far Right's Environmental Turn," *Dissent*, May 31, 2019, www.dissentmagazine.org/online_articles/the-european-far-rights-environmental-turn.

9. Alexander Ruser and Amanda Machin, "Nationalising the Climate: Is the European Far Right Turning Green?" *Green European Journal*, September 27, 2019, www.greeneuropeanjournal.eu/nationalising-the-climate-is-the-european-far-right-turning-green.

10. "Climate Refugees," Resource Library, National Geographic Society, March 28, 2019, www.nationalgeographic.org/encyclopedia/climate-refugees.

11. Such frameworks were often understood as sympathetic portrayals of peoples affected by colonial rule or social marginalization. See, for example, Edward E. Evans-Pritchard, *The Nuer: A Description of the Modes of Livelihood and Political Institutions of a Nilotic People* (Oxford: Clarendon Press, 1940). For critiques of this racialized mode of anthropological knowledge, see Denise Ferreira da Silva, *Toward a Global Idea of Race* (Minneapolis: University of Minnesota Press, 2007), 139–44; Mwenda Ntarangwi, *Reversed Gaze: An African Ethnography of American Anthropology* (Urbana: University of Illinois Press, 2010).

12. For a classic postcolonial critique of such gendered colonial rescue narratives, see Chandra Mohanty, "Under Western Eyes: Feminist Scholarship and Colonial Discourses," in *Colonial Discourse and Post-Colonial Theory: A Reader*, ed. Patrick Williams and Laura Chrisman (New York: Columbia University Press, 1989), 196–220.

13. Bina Desai, Justin Ginnetti, and Chloe Sydney, *No Matter of Choice: Displacement in a Changing Climate, Research Agenda and Call for Partners* (Geneva: IDMC, 2018), 1, 9.

14. Harsha Walia, *Undoing Border Imperialism* (Oakland: AK Press/Institute for Anarchist Studies, 2013), 20.

15. Yen Lê Espiritu, *Body Counts: The Vietnam War and Militarized Refuge(es)* (Berkeley: University of California Press, 2014), 13.

16. Mimi Thi Nguyen, *The Gift of Freedom: War, Debt, and Other Refugee Passages* (Durham, NC: Duke University Press, 2012), 2–3.

17. *Global Report on Internal Displacement 2018* (Geneva: IDMC, 2018), v.

18. *Global Report on Internal Displacement 2018*, 38.

19. Malini Ranganathan, "Thinking with Flint: Racial Liberalism and the Roots of an American Water Tragedy," *Capitalism Nature Socialism* 27, no. 3 (2016): 17–33.

20. A. Naomi Paik, *Rightlessness: Testimony and Redress in US Prison Camps since World War II* (Chapel Hill: University of North Carolina Press, 2016), 91.

21. Norman Myers and Jennifer Kent, *Environmental Exodus: An Emergent Crisis in the Global Arena* (Washington, DC: Climate Institute, 1995), 27.

22. Myers and Kent, *Environmental Exodus*, 35.

23. Myers and Kent, *Environmental Exodus*, 26, 31–32.

24. Myers and Kent, *Environmental Exodus*, 63.

25. See Paul Ehrlich and Anne Ehrlich, *The Population Bomb* (New York: Ballantine, 1968); Lester Brown and Hal Kane, *Full House: Reassessing the Earth's Population Capacity* (Washington, DC: Earthscan, 1995).

26. On Mayo's *Mother India* and visual stereotypes of Mohanty's figure of the third world woman, see Neel Ahuja, "Colonialism," in *Gender: Matter*, ed. Stacey Alaimo (New York: Macmillan, 2017), 247.

27. Nicholas De Genova, "Spectacles of Migrant 'Illegality': The Scene of Exclusion, the Obscene of Inclusion," *Ethnic and Racial Studies* 36, no. 7 (2013): 1181.

28. Walia, *Undoing Border Imperialism*, 7.

29. "The frame of 'refugee crisis' is largely a state- and media-generated response. I don't know that it emerges from any movements. And I think that is not a coincidence because the language of crisis meant that the crisis was somehow being experienced *by states*, and not *people*." Harsha Walia, interview by Neda Atanasoski and Christine Hong, *Critical Ethnic Studies* 6, no. 1 (Spring 2020), https://manifold.umn.edu/read/prefiguring-border-justice-interview-with-harsha-walia/section/96855cdc-5a01-45cd-9f1f-fc54b4ea6289. For Walia, the panics over migration invoke "the state itself as victim." Walia, *Undoing Border Imperialism*, 25.

30. Kelly Buchanan, *New Zealand: "Climate Change Refugee" Case Overview*, July 2015, www.loc.gov/law/help/climate-change-refugee/new-zealand.php.

31. "UN Landmark Case for People Displaced by Climate Change," January 20, 2020, www.amnesty.org/en/latest/news/2020/01/un-landmark-case-for-people-displaced-by-climate-change.

32. Melissa Godin, "Climate Refugees Cannot Be Forced Home, UN Panel Says in Landmark Ruling," *Time*, January 20, 2020, https://time.com/5768347/climate-refugees-un-ioane-teitiota/.

33. Sam O'Neill, "Amnesty International Expands Remit to Include Climate Change," *Times* (UK), April 30, 2019, www.thetimes.co.uk/article/amnesty-international-expands-remit-to-include-climate-change-w9zs38mmn.

34. Philip Alston, *Climate Change and Poverty: Report of the Special Rapporteur on Extreme Poverty and Human Rights*, UN Human Rights Council, 41st sess., June 24, 2019, https://undocs.org/pdf?symbol=en/A/HRC/41/39.

35. UN Human Rights Committee, "Views Adopted by the Committee under Article 5(4) of the Optional Protocol, Concerning Communication No. 2728/2016," CCPR/C/127/D/2727/2016, September 23, 2020.

36. Kenneth R. Weiss, "The Making of a Climate Refugee," *Foreign Policy*, January 28, 2015, https://foreignpolicy.com/2015/01/28/the-making-of-a-climate-refugee-kiribati-tarawa-teitiota.

37. Epeli Hau'ofa, "Our Sea of Islands," *Contemporary Pacific* 6, no. 1 (1994): 147–61 (quote on 151).

38. Cf. Achille Mbembe, "Necropolitics," trans. Libby Meintjes, *Public Culture* 15, no. 1 (2003): 11–40.

39. Tim McDonald, "The Man Who Would Be the First Climate Refugee," BBC News, November 5, 2015, www.bbc.com/news/world-asia-34674374.

40. Philip J. Deloria, *Playing Indian* (New Haven, CT: Yale University Press, 1998).

41. Renisa Mawani, *Across Oceans of Law: The* Komagata Maru *and Jurisdiction in the Time of Empire* (Durham, NC: Duke University Press, 2018).

42. Suvendrini Perera, "Oceanic Corpo-Geographies, Refugee Bodies, and the Making and Unmaking of Waters," *Feminist Review* 103 (2013): 58–79.

43. On climate change's queer temporalities, see Neel Ahuja, "Intimate Atmospheres: Queer Theory in a Time of Extinctions," *GLQ* 21, no. 2–3 (2015): 365–85.

44. Elizabeth DeLoughrey, *Allegories of the Anthropocene* (Durham, NC: Duke University Press, 2019), 128.

45. Veena Das, *Life and Words: Violence and the Descent into the Ordinary* (Berkeley: University of California Press, 2006).

46. See Rob Nixon, *Slow Violence and the Environmentalism of the Poor* (Cambridge, MA: Harvard University Press, 2011).

47. Aimee Bahng, *Migrant Futures: Decolonizing Speculation in Financial Times* (Durham, NC: Duke University Press, 2018), 11.

48. Refugee Appeal No. 72185 [2000] NZRSAA 335, August 10, 2000, www.nzlii.org/nz/cases/NZRSAA/2000/335.html.

49. Ruth Wilson Gilmore, *Golden Gulag: Prisons, Surplus, Opposition, and Crisis in Globalizing California* (Berkeley: University of California Press, 2007), 28.

50. See Nguyen, *The Gift of Freedom*.

51. See, for example, Laura Pulido, *Environmentalism and Economic Justice: Two Chicano Struggles in the Southwest* (Tucson: University of Arizona Press, 1996); Julie Sze, *Noxious New York: The Racial Politics of Urban Health and Environmental Justice* (Cambridge, MA: MIT Press, 2008); David Pellow, *Resisting Global Toxics: Transnational Movements for Environmental Justice* (Cambridge, MA: MIT Press, 2014); David Pellow, "Toward a Critical Environmental Justice Studies: Black Lives Matter as an Environmental Justice Challenge," *DuBois Review* 13, no. 6 (2016).

52. Tarak Barkawi and Mark Laffey, "The Postcolonial Moment in Security Studies," *Review of International Studies* 32 (2006): 350.

53. Key philosophical texts discuss this shift from biopower vested on the territorialized control of the human-as-species to such visions of ecologically integrated or networked control. See Michel Foucault, *Security, Territory, Population: Lectures at the Collège de France 1977–1978*, trans. Graham Burchell (Houndsmills: Palgrave, 2007), 4–6; Gilles Deleuze, "Postscript on the Societies of Control," *October* 59 (Winter 1992): 3–7.

54. Jasbir Puar, "Prognosis Time: Toward a Geopolitics of Affect, Debility, and Capacity," *Women and Performance* 19, no. 2 (2009): 161–72, esp. 164–66.

55. See Seb Franklin, *Control: Digitality as Cultural Logic* (Cambridge, MA: MIT Press, 2015).

56. IPCC AR5, Working Group II, *Climate Change 2014: Impacts, Adaptation, and Vulnerability*, pt. A, *Global and Sectoral Aspects* (Cambridge: Cambridge University Press, 2014), 767.

57. IPCC AR5, Working Group II, *Climate Change 2014*, 771.

58. Arturo Escobar, "Latin America at a Crossroads," *Cultural Studies* 24, no. 1 (2010): 1–65. The IPCC describes culture as a threatened resource for adaptation and resilience. IPCC AR5, Working Group II, *Climate Change 2014*, 762–66.

59. Kyle Powys White, Chris Powell, and Marie Shaeffer, "Indigenous Lessons about Sustainability Are Not Just for 'All Humanity,'" in *Sustainability: Approaches to Environmental Justice and Social Power*, ed. J. Sze (New York: NYU Press, 2018), 149–79.

60. David Pellow, *What Is Critical Environmental Justice?* (New York: Polity, 2018).

61. Laura Pulido, "A Critical Review of the Methodology of Environmental Racism Research," *Antipode* 28, no. 2 (1996): 142–59.

62. Julie Sze and Jonathan K. London, "Environmental Justice at the Crossroads," *Sociology Compass* 2, no. 4 (2008): 1332; Ranganathan, "Thinking with Flint"; Kyle Powys Whyte, "Indigenous Experience, Environmental Justice, and Settler Colonialism," *English Language Notes* 55, no. 1–2 (2017): 153–62.

63. Arun Saldanha, "Some Principles of Geocommunism," *Geocritique* (blog), July 2013, www.geocritique.org/arun-saldanha-some-principles-of-geocommunism/

64. Ruth Wilson Gilmore, *Golden Gulag: Prisons, Surplus, Opposition, and Crisis in Globalizing California* (Berkeley: University of California Press, 2007), 28.

65. Nikhil Singh, "Racial Formation in an Age of Permanent War," in *Racial Formation in the Twenty-First Century*, ed. Daniel Martinez HoSang, Oneka LaBennett, and Laura Pulido (Berkeley: University of California Press, 2012), 284.

66. Gilmore, *Golden Gulag*, 244.

Chapter Two

1. Yen Lê Espiritu, Lisa Lowe, and Lisa Yoneyama, "Transpacific Entanglements," in *Flashpoints for Asian American Studies*, ed. Cathy Schlund-Vials (New York: Fordham University Press, 2018), 175.

2. Barbara Ransby, *Making All Black Lives Matter: Reimagining Freedom in the Twenty-First Century* (Berkeley: University of California Press, 2018), 8.

3. This research is varied and includes a wide variety of perspectives on relationships between capitalism, anti-Black racism(s), and the state. See, among others, W. E. B. DuBois, "The Souls of White Folk," in *Darkmatter: Voices from within the Veil* (New York: Washington Square, 1920); C. L. R. James, *The Black Jacobins: Toussaint L'Overture and the San Domingo Revolution*, 2nd ed. (London: Secker and Warburg, 1938; New York: Vintage, 1989); W. E. B. DuBois, "Negroes and the Crisis of Capitalism in the US," *Monthly Review* 4, no. 12 (1953); Oliver Cromwell Cox, *Caste, Class, and Race: A Study in Social Dynamics* (New York: Monthly Review, 1959); Cedric Robinson, *Black Marxism: The Making of the Black Radical Tradition* (Chapel Hill: University of North Carolina Press, 1983); Adolph Reed, "The Black Urban Regime: Structural Origins and Constraints," *Comparative Urban and Community Research: An Annual Review* 1 (1988); Barbara Fields, "Slavery, Race, and Ideology in the United States of America," *New Left Review* 181 (1990): 95–118; Ruth Wilson Gilmore, *Golden Gulag: Prisons, Surplus, Crisis, and Opposition in Globalizing California* (Berkeley: University of California Press, 2007); Keeanga-Yahmatta Taylor, *From #Blacklivesmatter to Black Liberation* (Chicago: Haymarket, 2016).

4. See, as key reference points, Glen Coulthard, *Red Skins, White Masks: Rejecting the Colonial Politics of Recognition* (Minneapolis: University of Minnesota Press, 2014); Iyko Day, *Alien Capital: Asian Racialization and the Logic of Colonial Capitalism* (Durham, NC: Duke University Press, 2016); Patrick Wolfe, *Traces of History: Elementary Structures of Race* (London: Verso, 2016); Nikhil Pal Singh, *Race and America's Long War* (Berkeley: University of California Press, 2017).

5. See, most notably, Andre Gunder Frank, "Development and Underdevelopment in the New World: Smith and Marx vs. the Weberians," *Theory and Society* 2 (1975): 431–66; Immanuel Wallerstein, "The Rise and Future Demise of the World Capitalist Sys-

tem: Concepts for Comparative Analysis," *Comparative Studies in Society and History* 16 (1974): 387–415; L. S. Stavrianos, *Global Rift: The Third World Comes of Age* (New York: Morrow, 1981); Giovanni Arrighi, *The Long Twentieth Century* (London: Verso, 1994); Janet L. Abu-Laghod, *Before European Hegemony: The World System, A.D. 1250–1350* (Oxford: Oxford University Press, 1989); Walter Rodney, *How Europe Underdeveloped Africa* (Washington, DC: Howard University Press, 1974); Samir Amin, *Unequal Development: An Essay on the Social Formations of Peripheral Capitalism* (New York: Monthly Review, 1977); Samir Amin, *Eurocentrism* (New York: Monthly Review, 2009); Vijay Prashad, *The Darker Nations: A People's History of the Third World* (New York: New Press, 2007).

6. See, most notably, Silvia Federici, *Caliban and the Witch: Women, the Body, and Primitive Accumulation* (Brooklyn: Autonomedia, 2004).

7. Robinson, *Black Marxism*, esp. chap. 1.

8. See also Cheryl Harris, "Whiteness as Property," *Harvard Law Review* 106, no. 8 (1993): 1709–91.

9. One might reasonably object to Robinson's account by arguing that trade is not synonymous with capital. Distance trades, including trade in slaves, were central to precapitalist economies. See Abu-Laghod, *Before European Hegemony*. The historical forms of trade across the Mediterranean, Indian Ocean, and China in the eras prior to the rise of European overseas colonial networks have been the subject of extensive study by world system theorists and historians. As such, economic historians following Robert Brenner often argue that trade itself was not the engine that systemically formed the capitalist system but rather the enclosures and the expansion of capitalist property in land, which displaced the feudal landlord and enabled the bourgeoisie to dominate the emerging markets of Europe. See Brenner, "The Origins of Capitalist Development: A Critique of Neo-Smithian Marxism," *New Left Review* 104 (1977): 25–92. Such an argument would tend to center the history of Britain over the southern European merchant subcultures that created the Atlantic slave trade and that became central to Robinson's account of the rise of capitalism. It would also need to contend with Robinson's claim that despite the relatively small social groupings of the new merchant class, the spatial relations they developed by first expanding the manufacturing base and then causing monarchal states to invest in massive colonial enterprises helped to generate the systemic nature of colonial capitalism. Regardless, Robinson's point about the historical development of race is that the tendency toward racialized differentiation of labor and trade was already inherent to feudal production. It would be disingenuous to argue that property in land was not similarly racialized, especially given that European imperial justifications for land seizure in the Americas and justifications for property in persons (slavery) often coincided with dismissals of African and Indigenous sovereignty. See Lisa Lowe, *The Intimacies of Four Continents* (Durham, NC: Duke University Press, 2015); Cheryl Harris, "Whiteness as Property." Regardless of whether the property regime or trade practices contributed more to the transnational structure of capital in the colonial Atlantic, racially differentiated labor and property were, according to Robinson, an outgrowth of feudal forms of differentiation that constituted the social milieu of early capitalism.

10. Singh, *Race and America's Long War*.

11. Robinson, *Black Marxism*, 26.

12. Robert Vitalis, *America's Kingdom: Mythmaking on the Saudi Oil Frontier* (Stanford, CA: Stanford University Press, 2007), 119.

13. Taylor, *From #Blacklivesmatter to Black Liberation*, 205.

14. "Saudi Arabia Regularises Status of 40 Lakh Expatriate Workers," *Zee News*, July 17, 2013, https://zeenews.india.com/news/world/saudi-arabia-regularises-status-of-40-lakh-expatriate-workers_862797.html.

15. Bassam Za'za, "45 Indian Workers Sentenced for Unauthorised Labour Strike," *Gulf News*, February 24, 2008, https://gulfnews.com/uae/crime/45-indian-workers-sentenced-for-unauthorised-labour-strike-1.86282.

16. Mae Ngai, "The Architecture of Race in American Immigration Law: A Reexamination of the Immigration Act of 1924," *Journal of American History* 86, no. 1 (1999): 67–92.

17. Iyko Day, *Alien Capital: Asian Racialization and the Logic of Settler Colonial Capitalism* (Durham, NC: Duke University Press, 2016), 34. For Day, writing on the transition in the United States from dependency on enslaved Black labor to migrant Asian labor, "Domestic racial control served a broader logic of exclusion that is inherent in immigration restriction: to underscore and preserve Asians' alien status by creating policy that exploited the volatility of Asian presence. From the perspective of settler colonialism, we can build on this framework by clarifying the importance of spatial alienation (rather than Indigeneity) as a factor in the exploitation of a racialized labor force. In this light, a logic of exclusion is a *prerequisite* for the recruitment of alien labor, functioning either to produce an exclusive labor force in the case of African slaves or to render an Asian labor presence highly conditional to the demands of capital. Both are subject to forms of segregation, on either a national or international scale." On border militarization as the underside of the capitalist demand for labor, see Nicholas De Genova, "Spectacles of Migrant 'Illegality': The Scene of Exclusion, the Obscene of Inclusion," *Ethnic and Racial Studies* 36, no. 7 (2013): 1180–98.

18. Lowe, *The Intimacies of Four Continents*; Walton Look Lai, *Indentured Labor, Caribbean Sugar: Chinese and Indian Migrants to the British West Indies, 1838–1918* (Baltimore: Johns Hopkins University Press, 1993).

19. For examples in popular history and leftist accounts of empire, see, respectively, Daniel Yergin, *The Prize: The Epic Quest for Oil, Money, and Power* (New York: Simon and Schuster, 1991); David Harvey, *The New Imperialism* (Oxford: Oxford University Press, 2005).

20. Timothy Mitchell, *Carbon Democracy: Political Power in the Age of Oil* (London: Verso, 2011), 44–45.

21. Vinay Gidwani and Rajyashree Reddy, "The Afterlives of 'Waste': Notes from India for a Minor History of Capitalist Surplus," *Antipode* 43, no. 5 (2011): 1625–26.

22. Adam Smith, *An Inquiry into the Nature and Causes of the Wealth of Nations* (Amsterdam: MetaLibri, 2007), 437.

23. Gidwani and Reddy, "Afterlives of 'Waste,'" 1632–34.

24. Patrick Wolfe, "Land, Labor, and Difference: Elementary Structures of Race," *American Historical Review* 106, no. 3 (2001): 866–905; Iyko Day, *Alien Capital: Asian Racialization and the Logic of Settler Colonial Capitalism* (Durham, NC: Duke University Press, 2016), esp. 25–28.

25. Michael Watts, "Reflections on Circulation, Logistics, and the Frontiers of Capitalist Supply Chains," *Society and Space* 37, no. 5 (2019): 2042–49.

26. Mitchell, *Carbon Democracy*, 38–39. According to Mitchell, "Whereas the movement of coal tended to follow dendritic networks, with branches at each end but a single main channel, creating potential choke points at several junctures, oil flowed along networks that often had the properties of a grid, like an electricity network, where there is more than one possible path and the flow of energy can switch to avoid blockages or overcome breakdowns. These changes in the way forms of fossil energy were extracted, transported and used made energy networks less vulnerable to the political claims of those whose labour kept them running. Unlike the movement of coal, the flow of oil could not readily be assembled into a machine that enabled large numbers of people to exercise novel forms of political power."

27. Patrick Wolfe, "Land, Labor, and Difference: Elementary Structures of Race," *American Historical Review* 106, no. 3 (2001): 870.

28. Mitchell, *Carbon Democracy*, 209–10.

29. I. J. Seccombe and R. I. Lawless, "Foreign Worker Dependence in the Gulf, and the International Oil Companies, 1910–50," *International Migration Review* 20, no. 3 (1986): 563.

30. Mitchell, *Carbon Democracy*, 208.

31. Dai Jinhua, *After the Post-Cold War: The Future of Chinese History*, ed. Lisa Rofel (Durham, NC: Duke University Press, 2018), 17. Dai takes particular note of the sketch of Chinese economic dominance in Giovanni Arrighi, *Adam Smith in Beijing: Lineages of the Twenty-First Century* (London: Verso, 2007).

32. Richard Nixon, "Address to the Nation Outlining a New Economic Policy: 'The Challenge of Peace,'" August 15, 1971, American Presidency Project, www.presidency.ucsb.edu/documents/address-the-nation-outlining-new-economic-policy-the-challenge-peace.

33. Manolo Abella, "Asian Migrant Contract Workers in the Middle East," in *The Cambridge Survey of World Migration*, ed. R. Cohen (Cambridge: Cambridge University Press, 1995), 418–23; Andrzej Kapiszewski, "Arab versus Asian Migrant Workers in the GCC Countries," in *South Asian Migration to Gulf Countries: Histories, Policies, Development*, ed. Prakash C. Jain and Ginu Zacharia Oomen (London: Routledge, 2016), 46–70.

34. Prashad, *Darker Nations*, 187–88.

35. Prashad, *Darker Nations*, 187–88.

36. Radhika Desai, *Geopolitical Economy: After US Hegemony, Globalization, and Empire* (London: Pluto, 2013), 158–61; Prashad, *Darker Nations*, 188; Atif Kubursi and Salim Mansur, "The Political Economy of Middle Eastern Oil," in *Political Economy and the Changing Global Order*, ed. Richard Stubbs and Geoffrey R. D. Underhill (New York: St. Martin's Press, 1994), 313–27; David Spiro, *The Hidden Hand of American Hegemony: Petrodollar Recycling and International Markets*.

37. Jodi Kim, "Settler Modernity, Debt Imperialism, and the Necropolitics of the Promise," *Social Text* 135 (2018): 41–61.

38. David Harvey, "The 'New' Imperialism: Accumulation by Dispossession," *Socialist Register* 40 (2004): 74.

39. Saskia Sassen, "Women's Burden: Counter-Geographies of Globalization and the Feminization of Survival," *Journal of International Affairs* 53, no. 2 (2000): 503–24; Nasra M. Shah, "Gender and Labour Migration to the Gulf Countries," *Feminist Review* 77, no. 1 (2004): 183–85.

40. Alan Richards and John Waterbury, *A Political Economy of the Middle East*, 4th ed. (London: Routledge, 2013).

41. Radhika Desai, *Geopolitical Economy*.

42. Peter Beaumont, "The $18bn Arms Race Helping to Fuel Middle East Conflict," *Guardian*, April 23, 2015, www.theguardian.com/world/2015/apr/23/the-18bn-arms-race-middle-east-russia-iran-iraq-un.

43. In his study of Nixon's shift from the gold standard to a floating dollar as central to "the economic strategy of American empire," Michael Hudson describes the resulting inequality in debt obligations as a primary inequity in the global financial system. Although Hudson uses the term "strategy," he clarifies that the outcome was not initially envisioned by the Nixon administration. Hudson's Marxist account of financial empire as a strategy to overcome U.S. balance-of-payment difficulties and to maintain military dominance internationally is unique in its acceptance by some U.S. conservatives in government and think tanks (especially libertarians). They may be attracted by its emphasis on how Nixon's move enhanced U.S. power or by the possible suggestion that the gold standard provides a more rational system of valuation. After the book was published, Hudson was hired by the right-wing Hudson Institute and spent years there before returning to academia. See Michael Hudson, *Super Imperialism: The Economic Strategy of American Empire*, 2nd ed. (London: Pluto Press, 2003).

44. Phillip Connor, "International Migration: Key Findings from the U.S., Europe and the World," Pew Research Center, December 15, 2016, www.pewresearch.org/fact-tank/2016/12/15/international-migration-key-findings-from-the-u-s-europe-and-the-world.

45. Sabine Henning, "Overview of Global Trends in International Migration and Urbanization" (PowerPoint presentation, UN Expert Group Meeting on Sustainable Cities, Human Mobility and International Migration, New York, September 7, 2017), www.un.org/en/development/desa/population/events/pdf/expert/27/presentations/I/presentation-Henning-final.pdf.

46. Ting Ma, Rui Lu, Na Zhao, and Shih-Lung Shaw, "An Estimate of Rural Exodus in China Using Location-Aware Data," *PLoS ONE* 13, no. 7 (2018), https://doi.org/10.1371/journal.pone.0201458.

47. Kartik Roy, Clem Tisdell, and Mohammed Alauddin, "Rural-Urban Migration and Poverty in South Asia," *Journal of Contemporary Asia* 22 (1992): 57–72.

48. "Urbanization and Migration," Migration Data Portal, accessed February 2, 2020, https://migrationdataportal.org/themes/urbanisation-et-migration.

49. Nicholas De Genova, "The Borders of 'Europe' and the European Question," in Nicholas De Genova, ed., *The Borders of "Europe": Autonomy of Migration, Tactics of Bordering* (Durham, NC: Duke University Press, 2017), 1–36; Seth Holmes and Heide Castañeda, "Representing the 'European Refugee Crisis' in Germany and Beyond: Deservingness and Difference, Life and Death," *American Ethnologist* 43, no. 1 (2016): 12–24.

50. United Nations Human Settlements Programme, *World Cities Report 2016: Urbanization and Development: Emerging Futures* (Nairobi: UN-Habitat, 2016), chap. 4, http://wcr.unhabitat.org/wp-content/uploads/2017/03/Chapter4-WCR-2016.pdf.

51. Henning, "Overview of Global Trends."

52. Jennifer Burney and V. Ramanathan, "Recent Climate and Air Pollution Impacts on Indian Agriculture," *Proceedings of the National Academy of Sciences* 111, no. 46 (2014): 16319–24.

53. Marco Grasso, "Oily Politics: A Critical Assessment of the Oil and Gas Industry's Contribution to Climate Change," *Energy Research and Social Science* 50 (2019): 106–15.

54. See, most notably, Francis Fukuyama, *The End of History and the Last Man* (New York: Free Press, 1992). On postsocialism and postcolonialism, see Sharad Chari and Katherine Verdery, "Thinking between the Posts: Postcolonialism, Postsocialism, and Ethnography after the Cold War," *Comparative Studies in Society and History* 51, no. 1 (2009): 6–34.

55. Ruth Wilson Gilmore, "Race and Globalization," in *Geographies of Global Change: Remapping the World*, ed. R. J. Johnston, Peter Taylor, and Michael Watts, 2nd ed. (Malden: Blackwell, 2002), 265.

56. Thomas Friedman, *The World Is Flat: A Brief History of the Twenty-First Century* (New York: Farrar, Straus, and Giroux, 2005).

57. Harvey, "The 'New' Imperialism," 63–87.

58. Brenna Bhandar, *Colonial Lives of Property: Law, Land, and Racial Regimes of Ownership* (Durham, NC: Duke University Press, 2018).

59. Denise Ferreira da Silva and Paula Chakravartty, "Accumulation, Dispossession, and Debt: The Racial Logic of Global Capitalism—an Introduction," *American Quarterly* 64, no. 3 (2012): 364.

60. Robinson, *Black Marxism*, 26.

61. Robin Kelley, "What Is Racial Capitalism and Why Does It Matter?" (lecture, University of Washington, Seattle, WA, November 7, 2017), https://simpsoncenter.org/news/2017/10/robin-d-g-kelley-racial-capitalism-nov-7; Malini Ranganathan, "Thinking with Flint: American Liberalism and the Roots of an American Water Tragedy," *Capitalism Nature Socialism* 27, no. 3 (2016): 17–33.

62. Kasia Paprocki, "All That Is Solid Melts into the Bay: Anticipatory Ruination and Climate Change Adaptation," *Antipode* 51, no. 1 (2018): 295–315 (quote on 309).

63. Ariel Cohen, "Saudi Aramco IPO Hits $2 Trillion Mark amid Guarded Forecast," *Forbes*, December 18, 2019, www.forbes.com/sites/arielcohen/2019/12/18/saudi-aramco-ipo-hits-2-trillion-mark-but-forecast-still-guarded/#7ef3929942e6.

64. Frantz Fanon, "The Trials and Tribulations of National Consciousness," in *The Wretched of the Earth*, trans. Richard Philcox, rev. ed. (1963; New York: Grove, 2004), 97–144.

65. Kirk Hamilton, Quentin Wodon, Diego Barrot, and Ali Yedan, "Human Capital and the Wealth of Nations: Global Estimates and Trends," *The Changing Wealth of Nations: Building a Sustainable Future*, ed. Glenn-Marie Lange, Quentin Wodon, and Kevin Carey (Washington, DC: World Bank Group, 2018), 116.

66. Jason Moore, *Capitalism in the Web of Life: Ecology and the Accumulation of Capital* (London: Verso, 2015).

Chapter Three

1. Walton Look Lai, *Indentured Sugar, Caribbean Labor: Chinese and Indian Migrants to the British West Indies, 1838–1918*; Tejaswini Niranjana, *Mobilizing India: Women, Music, and Migration between India and Trinidad* (Durham, NC: Duke University Press, 2006); Lisa Lowe, *The Intimacies of Four Continents* (Durham, NC: Duke University Press, 2015).

2. Ginu Zacharia Oomen, "South Asian Migration to the GCC Countries: Emerging Trends and Challenges," in *South Asian Migration to Gulf Countries: History, Policies, Development*, ed. Prakash C. Jain and Ginu Zacharia Oomen (London: Routledge, 2016), 31.

3. Prakash C. Jain and Ginu Zacharia Oomen, introduction to *South Asian Migration to Gulf Countries*, 2.

4. This is the fourth major geographic formation of the South Asian diaspora, following the Indian Ocean merchant migrations, the indenture and Kangani systems in the Caribbean and Southeast Asia, and the postindependence migrations to the North.

5. Andrzej Kapiszewski, "Arab versus Asian Migrant Workers in the GCC Countries," in Jain and Oomen, *South Asian Migration to Gulf Countries*, 46–71; Zakir Hussain, "GCC's Immigration Policy in the Post-1990s: Contextualising South Asian Migration," in Jain and Oomen, *South Asian Migration to Gulf Countries*, 93–120.

6. Kapiszewski, "Arab versus Asian Migrant Workers," 56.

7. Oomen, "South Asian Migration to the GCC Countries," 22.

8. Rupananda Roy, "The Political Economy of Labour Migration from Bangladesh: Power, Politics, and Contestation" (PhD diss., University of Adelaide, 2016), 115–16.

9. Rita Afsar, "Revisiting the Saga of Bangladeshi Labour Migration to the Gulf States: Need for New Theoretical and Methodological Approaches," in Jain and Oomen, *South Asian Migration to Gulf Countries*, 159.

10. Roy, "Political Economy of Labour Migration from Bangladesh," 113–20, 130–31.

11. Oomen, "South Asian Migration to the GCC Countries," 24.

12. Kareem Fahim, "Saudi Arabia Encouraged Foreign Workers to Leave—and Is Struggling After So Many Did," *Washington Post*, February 2, 2019, https://www.washingtonpost.com/world/saudi-arabia-encouraged-foreign-workers-to-leave----and-is-struggling-after-so-many-did/2019/02/01/07e34e12-a548-11e8-ad6f-080770dcddc2_story.html

13. *Global Report on Internal Displacement 2018* (Geneva: IDMC, 2018), 7, www.internal-displacement.org/global-report/grid2018; "Dhaka: Improving Living Conditions for the Urban Poor" (Bangladesh Development Series Paper No. 17, World Bank, Washington, DC, 2007), xiii. IDMC notes that Bangladesh does not maintain systematic data on internal displacement.

14. Afsar, "Revisiting the Saga of Bangladeshi Labour Migration to the Gulf," 161–63.

15. *National Security and the Threat of Climate Change* (Alexandria, VA: CNA, 2007).

16. Marcus D. King and Ralph H. Espach, *Global Climate Change and State Stability* (Alexandria, VA: CNA 2009), 21.

17. Asma Khan Lone, "How Can Climate Change Trigger Conflict in South Asia?" *Foreign Policy*, November 20, 2015, https://foreignpolicy.com/2015/11/20/how-can-climate-change-trigger-conflict-in-south-asia.

18. *Climate Change and Security in Bangladesh: A Case Study* (Dhaka: Bangladesh Institute of International and Strategic Studies and Saferworld, 2009), 21.

19. Atiya Husain, "Retrieving the Religion in Racialization: A Critical Review," *Sociology Compass* 11, no. 3 (2017): e12507, https://doi/10.111/soc4.12507; Junaid Rana, "The Story of Islamophobia," *Souls* 9, no. 2 (2007): 148–61; Fatima El-Tayeb, *European Others: Queering Ethnicity in Postnational Europe* (Minneapolis: University of Minnesota Press, 2011), chap. 3. See also Etienne Balibar, "Is There a Neo-Racism?," in *Race, Nation, Class: Ambiguous Identities*, by Etienne Balibar and Immanuel Wallerstein (London: Verso, 1991), 17–28.

20. German Advisory Council on Global Change, *Climate Change as a Security Risk* (London: Earthscan, 2008), 3.

21. M. Sophia Newman, "Will Climate Change Spark Conflict in Bangladesh?" *Diplomat*, June 27, 2014, https://thediplomat.com/2014/06/will-climate-change-spark-conflict-in-bangladesh.

22. Gardiner Harris, "Borrowed Time on Disappearing Land," *New York Times*, March 28, 2014, www.nytimes.com/2014/03/29/world/asia/facing-rising-seas-bangladesh-confronts-the-consequences-of-climate-change.html.

23. Nicholas Kristof, "Swallowed by the Sea," *New York Times*, January 19, 2018, www.nytimes.com/2018/01/19/opinion/sunday/climate-change-bangladesh.html.

24. See Vijay Prashad, *The Karma of Brown Folk* (Minneapolis: University of Minnesota Press, 2000).

25. Kasia Paprocki, "All That Is Solid Melts into the Bay: Anticipatory Ruination and Climate Change Adaptation," *Antipode* 51, no. 1 (2019): 295–315.

26. George Naufal and Ismail Genc, *Expats and the Labor Force: The Story of the Gulf Cooperation Council Countries* (New York: Palgrave MacMillan, 2012), 21. Thanks to Naveeda Khan for discussing the IMF's change in designation with me.

27. Rita Afsar includes an example of crop loss leading to the decision to migrate to the Gulf in "Revisiting the Saga of Bangladeshi Labour Migration to the Gulf States," 166.

28. Harris, "Borrowed Time on Disappearing Land."

29. Government of the People's Republic of Bangladesh, *Bangladesh Climate Change Strategy and Action Plan* (Dhaka: Ministry of Environment and Forests, 2009), 17.

30. Nils Petter Gleditsch, "Armed Conflict and the Environment: A Critique of the Literature," *Journal of Peace Research* 35, no. 3 (1998): 381–400; Adano Wario Roba and Karen Witsenburg, *Surviving Pastoral Decline: Pastoral Sedentarisation, Natural Resource Management, and Livelihood Diversification in Marsabit District, Northern Kenya* (Amsterdam: Mellen, 2008); Ole Magnus Thiesen, Helge Holtermann, and Halvard Buhaug, "Climate Wars? Assessing the Claim That Drought Breeds Conflict," *International Security* 36, no. 3 (2011–12): 79–106; Nils Petter Gleditsch and Ragnhild Nordås, "Conflicting Messages? The IPCC on Conflict and Human Security," *Political Geography* 43 (2014): 82–90.

31. Peter Kropotkin, *Mutual Aid: A Factor of Evolution* (New York: McClure Phillips, 1909), www.gutenberg.org/cache/epub/4341/pg4341-images.html.

32. IPCC AR5, Working Group II, *Climate Change 2014: Impacts, Adaptation, Vulnerability*, pt. A, *Global and Sectoral Aspects* (Cambridge: Cambridge University Press, 2014), 758.

33. Don Belt, "The Coming Storm," *National Geographic*, May 2011, www.national geographic.com/magazine/2011/05/bangladesh.

34. Debjani Bhattacharya, *Empire and Ecology in the Bengal Delta: The Making of Calcutta* (Cambridge: Cambridge University Press, 2018).

35. Naveeda Khan, "The Death of Nature in the Era of Climate Change," in *Wording the World: Veena Das and Scenes of Inheritance*, ed. Roma Chaterji (New York: Fordham University Press, 2015), 292.

36. Khan, "Death of Nature in the Era of Climate Change," 292.

37. Dipesh Chakrabarty, "The Politics of Climate Change Is More Than the Politics of Capitalism," *Theory, Culture, and Society* 34, no. 2–3 (2017): 25–37.

38. Dipesh Chakrabarty, *Provincializing Europe* (Princeton, NJ: Princeton University Press, 2000), chap. 4.

39. Dipesh Chakrabarty, "The Climate of History: Four Theses," *Critical Inquiry* 35 (2009): 197–222.

40. Chakrabarty, "Climate of History," 222.

41. Chakrabarty, "Climate of History," 221–22.

42. Amitav Ghosh, *The Great Derangement* (Chicago: University of Chicago Press, 2016), 3–4.

43. Ghosh, *Great Derangement*, 5.

Chapter Four

1. Colin Kelley, Shahrzad Mohtadi, Mark A. Cane, Richard Seager, and Yochanan Kushnir, "Climate Change in the Fertile Crescent and Implications of the Recent Syrian Drought," *Proceedings of the National Academy of Sciences*, March 2, 2015, https://doi.org/10.1073/pnas.1421533112.

2. Craig Welch, "Climate Change Helped Spark Syrian War, Study Says," *National Geographic*, March 2, 2015, www.nationalgeographic.com/news/2015/3/150302-syria-war-climate-change-drought.

3. Robinson Meyer, "Does Climate Change Cause More War?," *Atlantic*, February 12, 2018, www.theatlantic.com/science/archive/2018/02/does-climate-change-cause-more-war/553040; Courtland Adams, Tobias Ide, Jon Barnett, and Adrien Detges, "Sampling Bias in Climate-Conflict Research," *Nature Climate Change* 8 (2018): 200–203, https://doi.org/10.1038/s41558-018-0068-2.

4. See "Water Conflict," Pacific Institute, 2019, www.worldwater.org/conflict.html; Peter H. Gleick, "Water, Drought, Climate Change, and Conflict in Syria," *Weather, Climate, and Society* 6, no. 3 (2014): 331–40, https://doi.org/10.1175/WCAS-D-13-00059.1.

5. General Charles H. Jacoby (Ret.), "The Biggest National Security Threat You Haven't Thought Of," *The Climate 25*, Weather Channel, accessed December 1, 2020, https://features.weather.com/climate25/project/general-charles-h-jacoby-ret.

6. David Livingstone, "Changing Climate, Human Evolution, and the Revival of Environmental Determinism," *Bulletin of the History of Medicine* 86, no. 4 (2012): 564–95; Steven Frenkel, "Geography, Empire, and Environmental Determinism," *Geographical Review* 82, no. 2 (1992): 143–53.

7. Robert Vitalis, *White World Order, Black Power Politics: The Birth of American International Relations* (Ithaca, NY: Cornell University Press, 2015), 193.

8. Iyko Day, *Alien Capital: Asian Racialization and the Logic of Settler Colonial Capitalism* (Durham, NC: Duke University Press, 2016).

9. See, most notably, Suzanna Sawyer and Arun Agrawal, "Environmental Orientalisms," *Cultural Critique* 45 (2000): 71–108. See also Donna Haraway, *Primate Visions: Gender, Race, and Nature in the World of Modern Science* (New York: Routledge, 1989); Ramachandra Guha, "American Environmentalism and Wilderness Preservation: A Third World Critique," *Environmental Ethics* 11, no. 1 (1989): 71–83; Londa Schiebinger, *Nature's Body: Gender in the Making of Modern Science* (Boston: Beacon Press, 1993); Mary Louise Pratt, *Imperial Eyes: Travel Writing and Transculturation* (London: Routledge, 1992).

10. Betsy Hartmann, "Rethinking Climate Refugees and Climate Conflict: Rhetoric, Reality, and the Politics of Policy Discourse," *Journal of International Development* 22 (2010): 234.

11. Raymond L. Bryant, "Beyond the Impasse: The Power of Political Ecology in Third World Environmental Research," *Area* 29, no. 1 (March 1997): 6–7.

12. Jan Selby and Clemens Hoffman, "Rethinking Climate Change, Conflict, and Security," *Geopolitics* 19, no. 4 (2014): 748.

13. David Mitchell and Sharon Snyder, "Narrative Prosthesis and the Materiality of Metaphor," *The Disability Studies Reader*, ed. Lennard Davis, 2nd ed. (London: Routledge, 2006), 205.

14. Rosemarie Garland-Thomson, "The Politics of Staring: Visual Rhetorics of Disability in Popular Photography," in *Disability Studies: Enabling the Humanities*, ed. Sharon Snyder, Brenda Jo Bruggeman, and Rosemarie Garland-Thomson (New York: Modern Language Association, 2002), 63.

15. For classic postcolonial feminist critiques of humanitarian representation, see Gayatri Chakravorty Spivak, "Can the Subaltern Speak?," in *Marxism and the Interpretation of Culture*, ed. Lawrence Grossberg and Cary Nelson (London: MacMillan, 1988), 271–313; Chandra Talpade Mohanty, "Under Western Eyes: Feminist Scholarship and Colonial Discourses," in *Colonial Discourse and Postcolonial Theory: A Reader*, ed. Patrick Williams and Laura Chrisman (New York: Columbia University Press, 1994), 196–220.

16. See, for example, Simon Goodman, Ala Sirriyeh, and Simon McMahon, "The Evolving (Re)categorisations of Refugees throughout the 'Migrant/Refugee Crisis,'" *Community and Applied Social Psychology* 27, no. 2 (2017): 105–14; Dennis Lichtenstein, Jenny Ritter, and Birte Fähnrich, "The Migrant Crisis in German Public Discourse," in *The Migrant Crisis: European Perspectives and National Discourses*, ed. Melani Barlai, Birte Fähnrich, Christina Griessler, and Markus Rhomberg (Münster: LIT Verlag, 2017), 107–26; Ju-Sung Lee and Adina Nerghes, "Refugee or Migrant Crisis? Labels, Perceived Agency, and Sentiment Polarity in Online Discussions," *Social Media and Society* (July–September 2018): 1–22; Seth Holmes and Heidi Castañeda, "Representing

the 'European Refugee Crisis' in Germany and Beyond," *American Ethnologist* 43, no. 1 (2016): 12–24.

17. Fatima El-Tayeb, *European Others: Queering Ethnicity in Postnational Europe* (Minneapolis: University of Minnesota Press, 2011); Camilla Hawthorne, "In Search of Black Italia: Notes on Race, Belonging, and Activism in the Black Mediterranean," *Transition* 123 (2017): 152–74.

18. Nicholas De Genova, "Migration and the Mobility of Labor," in *The Oxford Handbook of Karl Marx*, ed. Matt Vidal, Tony Smith, Tomás Rotta, and Paul Prew (Oxford: Oxford University Press, 2018), 1–19, Oxford Handbooks Online; Nicholas De Genova, "The Borders of 'Europe' and the European Question," introduction to *The Borders of "Europe": Autonomy of Migration, Tactics of Bordering* (Durham, NC: Duke University Press, 2017), 1–36.

19. Jessica Corbett, "'We Have to Get This Right': Historic Bill in the US House Would Create Specific Protections for Climate Refugees," Common Dreams, October 14, 2019, www.commondreams.org/news/2019/10/24/we-have-get-right-historic-bill-us-house-would-create-specific-protections-climate.

20. Rob Bailey and Gemma Green, "Should Europe Be Concerned about Climate Refugees?," *Newsweek*, May 18, 2016, www.newsweek.com/should-europe-be-concerned-about-climate-refugees-460661.

21. Tom Bawden, "Refugee Crisis: Is Climate Change Affecting Mass Migration?," *Independent*, September 7, 2015, www.independent.co.uk/news/world/refugee-crisis-climate-change-affecting-mass-migration-10490434.html.

22. Frank Biermann, "Migrant Crisis: 'If We Don't Stop Climate Change . . . What We See Right Now Is Just the Beginning,'" interview by Phil McKenna, *Inside Climate News*, September 14, 2015, https://insideclimatenews.org/news/13092015/migrant-crisis-syria-europe-climate-change.

23. Nick Watts, et. al. "The *Lancet* Countdown on Health and Climate Change: From 25 Years of Inaction to a Global Transformation for Public Health," *Lancet* 391:10120 (October 30, 2017).

24. John Wendle, "Syria's Climate Refugees," *Scientific American*, March 2016, 52–53.

25. Wendle, "Syria's Climate Refugees," 52.

26. Jan Selby, Omar Dahi, Christiane Fröhlich, and Mike Hulme, "Climate Change and the Syrian War Revisited," *Political Geography* 60 (2017): 238.

27. Lina Eklund and Darcy Thompson, "Differences in Resource Management Affects Drought Vulnerability across the Borders between Iraq, Syria, and Turkey," *Ecology and Society* 22, no. 4 (2017), http://dx.doi.org/10.5751/ES-09179-220409.

28. Wendle, "Syria's Climate Refugees," 52.

29. Wendle, "Syria's Climate Refugees," 54.

30. Wendle, "Syria's Climate Refugees," 55.

31. Katharina Nett and Lukas Rüttinger, *Insurgency, Terrorism, and Organised Crime in a Warming Climate: Analysing the Links between Climate Change and Non-state Armed Groups* (Berlin: Adelphi, 2016), 23–24.

32. Samar Yazbek, *The Crossing: My Journey to the Shattered Heart of Syria* (London: Ryder, 2015), 43, 51–52.

33. Yazbek, *Crossing*, 8–9.

34. Yazbek, *Crossing*, 5.

35. Marwan Hisham, *Brothers of the Gun: A Memoir of the Syrian War*, illus. Molly Crabapple (New York: One World, 2018), 74, 68.

36. Hisham, *Brothers of the Gun*, 83, 105–6.

37. Thomas Friedman, "The Revolution Fueled by Climate Change," *The Climate 25*, Weather Channel, accessed December 1, 2020, https://features.weather.com/climate25/project/thomas-friedman.

38. Farah Nasif, "In Syria, 'Everything Changed with the Drought,'" *The Climate 25*, Weather Channel, accessed December 1, 2020, https://features.weather.com/climate25/project/farah-nasif.

39. Parenti, *Tropic of Chaos*, 5.

40. Parenti, *Tropic of Chaos*, 99–100.

41. Andreas Malm, "Tahrir Submerged? Five Theses on Revolution in the Era of Climate Change," *Nature Capitalism Socialism* 25, no. 3 (2014): 29.

42. Malm, "Tahrir Submerged?," 30.

43. Andreas Malm, "Revolution in a Warming World: Lessons from the Russian to the Syrian Revolutions," *Socialist Register* 53 (2017), https://climateandcapitalism.com/2018/03/17/malm-revolutionary-strategy.

44. Jesse Ribot, "Cause and Response: Vulnerability and Climate in the Anthropocene," in *Global Agrarian Transformations*, vol. 1, *New Directions in Agrarian Political Economy*, ed. Madeleine Fairbairn, Jonathan Fox, S. Ryan Isakson, Michael Levien, Nancy Lee Peluso, Shahra Razavi, Ian Scoones, and Kalyanakrishnan "Shivi" Sivaramakrishnan (London: Routledge, 2016), 16.

45. On October 9, 2019, Turkey invaded the Rojava autonomous region of northern Syria, displacing 300,000 people and killing dozens.

46. Yassin al-Haj Saleh, "The Dark Path of Minority Politics," April 19, 2019, www.yassinhs.com/2019/04/19/the-dark-path-of-minority-politics/#fn1.

47. Andy Heintz, "Dissidents of the Left: In Conversation with Yassin al-Haj Saleh," August 28, 2018, www.yassinhs.com/2018/08/28/dissidents-of-the-left-in-conversation-with-yassin-al-haj-saleh.

48. Abdullah Öcalan, "Abdullah Öcalan on the Return to Social Ecology" [translated excerpt of *Beyond State, Power, and Violence*], Make Rojava Green Again, April 10, 2019, https://makerojavagreenagain.org/2019/04/10/abdullah-ocalan-on-the-return-to-social-ecology.

49. *Make Rojava Green Again: Building an Ecological Society* (London: Internationalist Commune of Rojava/Dog Section Press, 2018).

Conclusion

1. Alejandra Mejía, "1.5 Gen Testimonios: My Family's Journey," Migrant Roots Media, March 5, 2019, www.migrantrootsmedia.org/articles/2019/3/5/15-generation-migrant-roots-un-testimonio-part-i-alejandra-meja.

2. Macarena Gomez-Barris, *The Extractive Zone: Social Ecologies and Decolonial Perspectives* (Durham, NC: Duke University Press, 2017), xviii–xix.

3. Ting Ma, Rui Lu, Na Zhao, and Shih-Lung Shaw, "An Estimate of Rural Exodus in China Using Location-Aware Data," *PLoS ONE*, July 31, 2018, https://journals.plos.org/plosone/article?id=10.1371/journal.pone.0201458.

4. "Social Contagion: Microbiological Class War in China," *Chuǎng* (blog), February 26, 2020, http://chuangcn.org/2020/02/social-contagion; Cihan Aksan and Jon Bailes, "One Question: COVID-19 and Capitalism," *State of Nature* (blog), March 27, 2020, http://stateofnatureblog.com/one-question-covid19-coronavirus-capitalism.

5. Bruce Braun, "Thinking the City through SARS: Bodies, Topologies, Politics," in *Networked Disease: Emerging Infections in the Global City*, ed. S. Harris Ali and Roger Keil (Oxford: Wiley-Blackwell, 2008), 250–66.

6. For discussions of how anti-Black labor exploitation is innovated in particular ways through later models of Asiatic labor accumulation, see Iyko Day, *Alien Capital: Asian Racialization and the Logics of Settler Colonial Capitalism* (Durham, NC: Duke University Press, 2016). Because Day's book categorizes Asiatic labor regimes within the same "alien" category as enslaved African Americans, it is necessary to complement discussion of such similarities with contextualization of the innovations of finance and distributed production that accompany twentieth- and twenty-first-century Asian diasporas.

7. See Candace Fujikane and Jonathan Y. Okamura, eds., *Asian Settler Colonialism: From Local Governance to the Habits of Everyday Life in Hawai'i* (Honolulu: University of Hawai'i Press, 2008).

8. Ruth Wilson Gilmore, *Golden Gulag: Prisons, Surplus, Crisis, and Opposition in Globalizing California* (Berkeley: University of California Press, 2008), 244.

9. On the idealism of posthumanist critique and the problem of anthropomorphizing the human, see Neel Ahuja, *Bioinsecurities: Disease Interventions, Empire, and the Government of Species* (Durham, NC: Duke University Press, 2016), viii, 8.

10. Etienne Balibar, "Is There a 'Neo-racism'?," in Etienne Balibar and Immanuel Wallerstein, *Race, Nation, Class: Ambiguous Identities* (London: Verso, 1991), 17–28.

11. See Bhishnupriya Ghosh, *Global Icons: Apertures to the Popular* (Durham, NC: Duke University Press, 2011).

Index

Page numbers appearing in italic type refer to pages containing illustrations.

MIX
Paper from
responsible sources
FSC® C008955
FSC
www.fsc.org